Pfaffian Systems, k-Symplectic Systems

Pfaffian Systems,
k-Symplectic Systems

by

Azzouz Awane
Université Hassan II,
Faculté des Sciences,
Casablanca, Maroc

and

Michel Goze
Université de Haute Alsace,
Faculté des Sciences et Techniques,
Mulhouse, France

Springer-Science+Business Media, B.V.

A C.I.P. Catalogue record for this book is available from the Library of Congress.

ISBN 978-90-481-5486-9   ISBN 978-94-015-9526-1 (eBook)
DOI 10.1007/978-94-015-9526-1

Printed on acid-free paper

Contents

Introduction

The theory of foliations and contact forms have experienced such great development recently that it is natural they have implications in the field of mechanics. They form part of the framework of what Jean Dieudonné calls "Elie Cartan's great theory of the Pfaffian systems", and which even nowadays is still far from being exhausted. The major reference work is without any doubt that of Elie Cartan on Pfaffian systems with five variables. In it one discovers there the bases of an algebraic classification of these systems, their methods of reduction, and the highlighting of the first fundamental invariants. This work opens to us, even today, a colossal field of investigation and the mystery of a ternary form containing the differential invariants of the systems with five variables always deligthts anyone who wishes to find out about them.

One of the goals of this memorandum is to present this work of Cartan - which was treated even more analytically by Goursat in its lectures on Pfaffian systems - in order to expound the classifications currently known. The theory of foliations and contact forms appear in the study of completely integrable Pfaffian systems of rank one. In each of these situations there is a local model described either by Frobenius' theorem, or by Darboux' theorem. It is this type of theorem which it would be desirable to have for a non-integrable Pfaffian system which may also be of rank greater than one. We know that such a model is insufficient for describing events more general than can met in statistical mechanics. One possible approach lies in the description of the associated dynamic systems not by an exterior equation alone by a *system* of exterior equations. This original step has already made it possible to recover the equations of the statistical mechanics suggested by Nambu. One of the goals of this work is to lay down the bases of a multi-symplectic mechanics.

The first part of this work is of a purely algebraic nature. It presents to it the concepts of exterior forms and systems of exterior forms. A particular

study is devoted to the case of the exterior systems defined by equations of
the symplectic type, and to those groups which leave this structure invariant . Chapter 4 and 5 are devoted to the Pfaffian systems. In them are
defined Cartan's invariants (the class of a system, characteristic spaces, associated derived systems, *etc*), and one again finds the traditional theorems
of Frobenius and Darboux. This latter theorem is approached by studing
the maximal integral manifolds of a Pfaffian equation not necessarily integrable, and doing this makes it possible rather quickly to find Darboux'
theorem "with parameters" quoted by Goursat. The known classifications
of the Pfaffian systems are also presented, these classifications being given
up to local isomorphism.

These chapters culminate in the systems with five variables and their char-
acteristics: there is an infinity of local models - which does not appear in
lower dimension (note that all the systems studied do not have singularities).
Cartan's ternary form giving the differential invariants of these systems is
reconsidered, and it is presented with a reduction inspired by the approach
which Goursat gives in his book to some Pfaffian systems.

The last part of this work is devoted to the differential study of multi-
symplectic systems. The classification of the systems of exterior forms of
degree 2 shows, here also, that there exists an infinity of models (chapter
3). The conditions for the existence of solutions of maximum dimension
make it possible to describe one particular system, that corresponding to
the k-symplectic systems. These latter thus seem to be models of systems
of exterior 2-forms of maximum rank. The interest in embarking upon their
study is obvious within this framework. One starts by defining the concepts
of Hamiltonian dynamic systems and Poisson brackets. One then describes
the (simplest) differential affine manifolds provided with such a structure,
and one establishes the link with the models of statistical mechanics proposed by Nambu. In all this final part use is made of the traditional concepts
of differential geometry (connection, G-structure, reduction, cohomology)
without giving any detailled reconsideration of their definitions.

This work is the result of a collaboration between the University of Haute
Alsace and the Faculty of Sciences Ben Msik of Casablanca, with the con-
tunued financial assistance of the Integrated Action (AI 809-94).

We wish to thank Professor Mabrouk Benhamou for carefully reading the manuscript and making many constructive comments.

Casablanca, Colmar
October 11, 1999

Azzouz AWANE, Michel GOZE

Chapter 1

EXTERIOR FORMS

1.1 Exterior algebra

1.1.1 The dual space

Let $\mathbb{K}$ be a commutative field and let E be an n-dimensional vector space over $\mathbb{K}$. The dual E^* of E is the vector space of linear forms on E. If $v \in E$ and $f \in E^*$, the value of f on v will be denoted by $\langle v, f \rangle$.

Let $\{e_1, \cdots, e_n\}$ be a basis of E. Let us denote by $e^i \in E^*$ the linear form on E defined by:

$$ e^i : e_j \longmapsto \langle e_j, e^i \rangle = \delta^i_j = \left\{ \begin{array}{ll} 1 & \text{if} \quad i = j, \\ 0 & \text{if} \quad i \neq j. \end{array} \right. $$

Then $\{e^1, \cdots, e^n\}$ is a basis of E^*, called the dual basis of $\{e_1, \cdots, e_n\}$. We can note that the isomorphism

$$ F : E \longrightarrow E^* $$

defined by $F(e_i) = e^i$ is not canonical. In fact, it depends on the choice of the basis of E. For example, let us consider the new basis $\{v_1, ..., v_n\}$ given by $v_1 = ae_1, v_i = e_i, i \neq 1, a \neq 0$. If F' is the isomorphism between E and E^* given by $F'(v_i) = v^i$ where $\{v^1, ..., v^n\}$ is the dual basis of $\{v_1, ..., v_n\}$, then $F'(e_1) = a^{-1}v^1 = a^{-2}e^1$ and $F' \neq F$ whenever $a^2 \neq 1$. On the other hand, we can verify that the spaces E and E^{**} are canonically isomorphic, E^{**} being the dual of E^*.

We can see that the tensor product $E \otimes E^*$ can be identified to the space of all linear transformations of E. In fact, if $v \in E$, and $f \in E^*$, we put

$$ (v \otimes f)(w) = \langle w, f \rangle v $$

for all $w \in E$. Then $v \otimes f$ defines a linear transformation of E, and by linearity we can extend it to all elements of $E \otimes E^*$. Conversely, if we have an endomorphism of E having (a_{ij}) as its matrix with respect to the basis $\{e_1, .., e_n\}$ of E, we can write this endomorphism $\sum a_{ij} e_i \otimes e^j$.

Similarly, we can see that the tensor product $E^* \otimes E^*$ corresponds to the space of bilinear forms on E. We can extend this study of tensor products of E and E^*. This leads to the general study of the tensor algebra

$$\begin{aligned} T(E) \;=\; & \mathbb{K} + E + E^* + E \otimes E + E \otimes E^* + E^* \otimes E + E^* \otimes E^* \\ & + E \otimes E \otimes E + ... \end{aligned}$$

1.1.2   The algebra of contravariant tensors

Consider the subalgebra $C(E)$ of $T(E)$ given by

$$\begin{aligned} C(E) \;=\; & \mathbb{K} + E + E \otimes E + E \otimes E \otimes E + ... \\ =\; & C^0(E) + C^1(E) + C^2(E) + ... \;, \end{aligned}$$

where $C^p(E) = E \otimes E \otimes ... \otimes E$ (p–times). An element of $C^p(E)$ is called a p-contravariant tensor and $C(E)$ is called the algebra of contravariant tensors of E. Similarly, we can define the subalgebra $C(E^*)$ of $T(E)$. It is the algebra of covariant tensors of E. There exists a natural identification to $C^p(E^*)$ with $(C^p(E))^*$. In fact, we define a pairing of $C(E)$ and $C(E^*)$ into $\mathbb{K}$ by putting:

$$\langle v, \gamma \rangle = 0 \quad \text{if } v \in C^p(E) \text{ and } \gamma \in C^q(E^*) \text{ with } p \neq q,$$

and, if $p = q$, we write

$$\begin{aligned} v \;&=\; \sum v^{i_1...i_p} e_{i_1} \otimes ... \otimes e_{i_p}, \\ \gamma \;&=\; \sum \gamma_{j_1...j_p} e^{i_1} \otimes ... \otimes e^{j_p} \end{aligned}$$

and we put

$$\langle v, \gamma \rangle = \sum v^{i_1...i_p} \gamma_{i_1...i_p}.$$

This gives the seemed identification.

1.1.3   The exterior algebra of E

The permutation group $\mathfrak{S}_p$ acts on the linear space $T^p(E)$. If σ is a permutation and $v_1 \otimes ... \otimes v_p \in T^p(E)$, we define

$$\sigma(v_1 \otimes ... \otimes v_p) = v_{\sigma(1)} \otimes ... \otimes v_{\sigma(p)}$$

and extend by linearity to all of $T^p(E)$.

Definition 1.1 *A tensor $t \in T^p(E)$ is called antisymmetric if it satisfies $\sigma(t) = \text{sgn}(\sigma)t$, where $\text{sgn}(\sigma) = +1$ if σ is even, and -1 if σ is odd.*

We note $\Lambda^p(E)$ the space of antisymmetric p-tensors on E. There is a projection

$$T^p(E) \rightarrow \Lambda^p(E)$$

given by

$$\mathcal{A}_p(t) = \frac{1}{p!} \sum_{\sigma \in S_p} (\text{sgn}(\sigma))\sigma(t).$$

Now, consider the space $\Lambda(E)$ given by

$$\Lambda(E) = \mathbb{K} + E + \Lambda^2(E) + \Lambda^3(E) + \dots .$$

In this space we define the product $\wedge$ by:

$$v \wedge w = \mathcal{A}_{p+q}(v \otimes w)$$

with $v \in \Lambda^p(E)$ and $w \in \Lambda^q(E)$. Thus $v \wedge w \in \Lambda^{p+q}(E)$ and

$$v \wedge w = (-1)^{pq} w \wedge v.$$

This product provides $\Lambda(E)$ with the structure of an associative algebra, called the exterior algebra of E.

We can carry out the same construction from the vector space E^*, and define the exterior algebra $\Lambda(E^*)$ of E^*.

1.1.4 The algebra of covariant tensors

Consider the subalgebra $T(E^*)$ of $T(E)$ given by

$$
\begin{aligned}
T(E^*) &= \mathbb{K} + E^* + E^* \otimes E^* + E^* \otimes E^* \otimes E^* + \dots \\
&= T^0(E^*) + T^1(E^*) + T^2(E^*) + \dots ,
\end{aligned}
$$

where $T^p(E^*) = E^* \otimes E^* \otimes \dots \otimes E^*$ (p-times). An element of $T^p(E^*)$ is called a p-covariant tensor and $T(E^*)$ is called the algebra of covariant tensors of E. We can define the exterior algebra $\Lambda^p(E^*)$ as in the previous section, considering this algebra as the exterior algebra of the vector space E^*. It is this object which will interest us hereafter. For this we will define the structure of this algebra directly.

1.2   The exterior algebra $\Lambda(E^*)$

Throughout this chapter we consider the real numbers as the base field
($\mathbb{K} = \mathbb{R}$).

1.2.1   Exterior p-forms on E

Definition 1.2 *A p-exterior form on E is a mapping*

$$\alpha:(v_1,\cdots,v_p) \in E^p \longmapsto \alpha(v_1,\cdots,v_p) \in \mathbb{R}$$

satisfying:

$$\alpha(v_1,\cdots,av_i+bw_i,\cdots,v_p) = a\alpha(v_1,\cdots,v_i,\cdots,v_p) + b\alpha(v_1,\cdots,w_i,\cdots,v_p)$$

for all $a,b \in \mathbb{R}, v_j, w_i \in E$, and

$$\alpha(v_1,\cdots,v_p) = 0$$

as soon as $j \leq p$ such that $v_{j-1} = v_j$.

Proposition 1.1 *If α is a p-exterior form on E, then it satisfies:*

$$\alpha(v_1,\cdots,v_i,v_{i+1},\cdots,v_p) = -\alpha(v_1,\cdots,v_{i+1},v_i,\cdots,v_p)$$

for all i.

Proof. In fact,

$$
\begin{aligned}
\alpha(v_1,\cdots,v_i,v_{i+1},\cdots,v_p) &= \alpha(v_1,\cdots,v_i,v_{i+1},\cdots,v_p) \\
&\quad + \alpha(v_1,\cdots,v_{i+1},v_{i+1},\cdots,v_p) \\
&= \alpha(v_1,\cdots,v_i+v_{i+1},v_{i+1},\cdots,v_p) \\
&= \alpha(v_1,\cdots,v_i+v_{i+1},v_{i+1},\cdots,v_p) \\
&\quad -\alpha(v_1,\cdots,v_i+v_{i+1},v_i+v_{i+1},\cdots,v_p) \\
&= -\alpha(v_1,\cdots,v_i+v_{i+1},v_i,\cdots,v_p) \\
&= -\alpha(v_1,\cdots,v_{i+1},v_i,\cdots,v_p).
\end{aligned}
$$

We deduce that
$$\alpha(v_1,\cdots,v_p) = 0$$

when v_i and v_j are equal or proportional. Then every p-exterior form α on E is zero when $p > \dim E = n$. ∎

It is easy to see that the p-exterior forms on E are the elements of $\Lambda^p(E^*)$. Then, the exterior algebra

$$\Lambda(E^*) = \bigoplus_0^n \Lambda^p(E^*),$$

with $\Lambda^0(E^*) = \mathbb{R}$, is the space of exterior forms on E.

1.2.2 Grassmann product of linear forms

We will define directly the product of the exterior algebra $\Lambda(E^*)$ in the particular case of linear forms. Let us consider two linear forms f^1 and f^2 on the vector space E. We define the 2−exterior form, denoted $f^1 \wedge f^2$ and called the *Grassmann product* of f^1 and f^2, by putting:

$$f^1 \wedge f^2(X,Y) = f^1(X)f^2(Y) - f^1(Y)f^2(X),$$

for all $X, Y \in E$. This product is non commutative and satisfies:

$$f^1 \wedge f^2 = -f^2 \wedge f^1.$$

We can also define this product for a large number of factors.

Definition 1.3 *Let* $f^1, f^2, \cdots, f^p$ *be linear forms of* E^*. *The Grassmann product of these forms is the p-exterior form, denoted* $f^1 \wedge f^2 \wedge \cdots \wedge f^p$ *and defined by:*

$$(f^1 \wedge f^2 \wedge \cdots \wedge f^p)(X_1, \cdots, X_p) = \det \begin{pmatrix} f^1(X_1) & f^2(X_1) & \cdots & f^p(X_1) \\ f^1(X_2) & f^2(X_2) & \cdots & f^p(X_2) \\ \vdots & \vdots & \vdots & \vdots \\ f^1(X_p) & f^2(X_p) & \cdots & f^p(X_p) \end{pmatrix}$$

for all $X_1, ..., X_p \in E$, *where* $\det(M)$ *stands for the determinant of matrix* M.

The fact that $f^1 \wedge f^2 \wedge ... \wedge f^p$ is a p-exterior form results from the properties of the determinant.

Proposition 1.2 *Let* $f^1, \cdots, f^p$ *be linear forms on* E. *These forms are linearly dependent on* E^* *if and only if the Grassmann product satisfies*

$$f^1 \wedge f^2 \wedge \cdots \wedge f^p = 0.$$

Proof. Let us suppose that the linear forms $f^1, \ldots, f^p$ are linearly independent and let us consider the vectors $X_1, \cdots, X_p$ satisfying $f^i(X_j) = \delta^i_j$. Thus the matrix $(f^i(X_j))$ is the identity matrix. Then

$$f^1 \wedge f^2 \wedge \cdots \wedge f^p(X_1, \cdots, X_p) = 1$$

and

$$f^1 \wedge f^2 \wedge \cdots \wedge f^p \neq 0.$$

Conversely, if the linear forms $f^1, \cdots, f^p$ are linearly dependent in E^* then for all vectors $(X_1, \cdots, X_p)$ the rank of the matrix $(f^i(X_j))$ is less than p and $\det(f^i(X_j)) = 0$. $\blacksquare$

1.2.3   Exterior product

In the preceding paragraph we have defined the product on the exterior algebra. We take again here this definition in the case of the algebra $\Lambda(E^*)$.

Definition 1.4 *Let α be a p-exterior form and β a q-exterior form E. The exterior product $\alpha \wedge \beta$ of the exterior forms α and β, is the $(p+q)$-exterior form on E, defined by:*

$$\alpha \wedge \beta(X_1, \cdots X_{p+q}) =$$

$$\frac{1}{p!q!} \sum_{\sigma \in \mathfrak{S}_{p+q}} \epsilon(\sigma)\alpha(X_{\sigma(1)}, \cdots, X_{\sigma(p)})\beta(X_{\sigma(p+1)}, \cdots, X_{\sigma(p+q)}),$$

where $\mathfrak{S}_{p+q}$ is the permutation group of $p+q$ letters and $\epsilon(\sigma) = \mathrm{sgn}(\sigma)$ the signature of the permutation σ.

According to the previous section, it is easy to see that $\alpha \wedge \beta$ is a $(p+q)$-exterior form.

Remark 1 *The Grassmann product of linear forms is nothing other than the exterior product of these forms.*

Since

$$\alpha \wedge \beta = (-1)^{pq}\beta \wedge \alpha$$

for all $\alpha \in \bigwedge^p(E^*)$ and $\beta \in \bigwedge^q(E^*)$, we deduce that $\bigwedge(E^*)$ is an associative non-commutative algebra.

We are now able to determine the dimension of spaces $\bigwedge^p(E^*)$.

Proposition 1.3 *Let $\{f^1, \cdots, f^n\}$ be a basis of E^*. Then the p-exterior forms*

$$f^{i_1} \wedge \cdots \wedge f^{i_p}$$

with $i_j \in \{1, \cdots, n\}$ and $i_1 < i_2 < \cdots < i_p$ give a basis of the vector space $\bigwedge^p(E^*)$. Then

$$\dim \bigwedge{}^p E^* = \binom{n}{p} \quad (p \le n)$$

and $\dim \bigwedge(E^*) = 2^n$.

Proof. Let $\{X_1, \cdots, X_n\}$ be a basis of E such that $\{f^1, \cdots, f^n\}$ be its dual basis. Let α be a p-exterior form on E. We set

$$a_{i_1 \cdots i_p} = \alpha(X_{i_1}, \cdots, X_{i_p}).$$

We then have

$$\alpha = \sum a_{i_1 \cdots i_p} f^{i_1} \wedge \cdots \wedge f^{i_p}.$$

This shows that the family $\{f^{i_1} \wedge \cdots \wedge f^{i_p}\}$ generates the space $\bigwedge^p(E^*)$.

The linear independence of these forms results from the previous propositions. ∎

1.3 The graded algebra $\bigwedge(E^*)$

In order to simplify the notation, $\bigwedge(E^*)$ will be denoted by $\bigwedge E^*$. We have just seen that the graded vector space

$$\bigwedge E^* = \bigoplus_0^n \bigwedge{}^p E^*$$

is provided with a non commutative associative algebra of dimension 2^n. Each element α of $\bigwedge E^*$ is written on the form:

$$\alpha = \alpha_0 + \alpha_1 + \cdots + \alpha_n \;, \quad \alpha_i \in \bigwedge{}^i E^*$$

with $\bigwedge^0 E^* = \mathbb{R}$. The i-th exterior form α_i is called the *homogeneous component of degree i of α*.

1.3.1   The interior product

Definition 1.5 *Let $\alpha \in \bigwedge^p E^*$ be a p-exterior form on E, $p \geq 1$, and let X be a vector belonging to E. The interior product of X and α is the $(p-1)$-exterior form, denoted $i(X)\alpha$ (or $X \rfloor \alpha$), defined by:*

$$i(X)\alpha(X_1, \cdots, X_{p-1}) = \alpha(X, X_1, \cdots, X_p)$$

for all $X_1, \cdots, X_{p-1} \in E$.

The mapping

$$i(X) : \alpha \longmapsto i(X)\alpha$$

from $\bigwedge^p E^*$ into $\bigwedge^{p-1} E^*$ is linear. We can extend this mapping to an endomorphism of $\bigwedge E^*$ by setting

$$i(X)\alpha = 0, \text{ if } \alpha = a \in \mathbb{R}$$

and

$$i(X)\alpha = \sum_{i=1}^{n} i(X)\alpha_i,$$

where $\alpha = \alpha_0 + \alpha_1 + \cdots + \alpha_n$ is the quantity defined above.

Proposition 1.4 *Let X and Y be two vectors belonging to E. Then we have*

1. $i(X + Y) = i(X) + i(Y)$,

2. $i(aX) = ai(X)$, $a \in \mathbb{R}$,

3. $i(X)i(Y) + i(Y)i(X) = 0$,

4. $i(X)i(X) = 0$.

It is easy to prove these identities.

Proposition 1.5 *Let α (respectively β) be a $p-$exterior form (respectively a $q-$exterior form) on E. For every vector $X \in E$, we have:*

$$i(X)(\alpha \wedge \beta) = (i(X)\alpha) \wedge \beta + (-1)^p \alpha \wedge (i(X)\beta).$$

Proof. Let us consider $p + q$ independent vectors $\{e_1, \cdots, e_{p+q}\}$ (if $p + q$ is strictly greater than n, the previous identity is trivial) and let us assume that $X = e_1$. We have

$$i(X)(\alpha \wedge \beta)(e_2, \cdots, e_{p+q}) = (\alpha \wedge \beta)(e_1, \cdots, e_{p+q})$$

$$= \tfrac{1}{p!q!} \sum_{s \in S_{p+q}} \epsilon(s)\alpha(e_{s(1)}, \cdots, e_{s(p)})\beta(e_{s(p+1)}, \cdots, e_{s(p+q)}).$$

The permutation group $\mathfrak{S}_{p+q}$ is a disjoint union of S_1 constituted of permutations of which the pre-image of 1 is less or equal to p, and of the set S_2 of permutations of which the pre-image of 1 is strictly greater than p. We have:

$$\frac{1}{p!q!} \sum_{s \in S_1} \epsilon(s)\alpha(e_{s(1)}, \cdots, e_{s(p)})\beta(e_{s(p+1)}, \cdots, e_{s(p+q)}) = (i(X)\alpha) \wedge \beta$$

and

$$\frac{1}{p!q!} \sum_{s \in S_2} \epsilon(s)\alpha(e_{s(1)}, \cdots, e_{s(p)})\beta(e_{s(p+1)}, \cdots, e_{s(p+q)}) = (-1)^p \alpha \wedge (i(X)\beta).$$

This gives the result desired. ∎

Consider the one-dimensional vector spaces $\bigwedge^n E$ and $\bigwedge^n E^*$, where $n = \dim E$. These spaces are dual:

$$\bigwedge{}^n E^* = (\bigwedge{}^n E)^*.$$

Let e and e^* be the dual bases of $\bigwedge^n E$ and $\bigwedge^n E^*$.

Proposition 1.6 *The linear mapping*

$$\perp : \bigwedge{}^r E \longrightarrow \bigwedge{}^{n-r} E^*$$

defined by

$$\perp (X) = i(X)e^*$$

is an isomorphism of $\bigwedge^r E$ onto $\bigwedge^{n-r} E^*$. This mapping is unique up to a factor only.

Proof. We choose dual bases $\{e_1, ..., e_n\}$ and $\{\alpha_1 = e^1, ..., \alpha_n = e^n\}$ so that $e_1 \wedge ... \wedge e_n = e$ and $e^1 \wedge ... \wedge e^n = e^*$, then

$$i(e_1 \wedge ... \wedge e_n)(e^1 \wedge ... \wedge e^n) = 1$$

and

$$i(e_{i_1} \wedge ... \wedge e_{i_p})(e^{j_1} \wedge ... \wedge e^{j_r}) = \begin{cases} \pm e^{s_1} \wedge ... \wedge e^{s_{r-p}} \text{ if } \{i_1, ..., i_p\} \subset \{j_1, .., j_r\} \\ \\ 0 \text{ if not} \end{cases}$$

where $s_1, ..., s_{r-p}$ are the complementary set of indices in $j_1, .., j_r$ to $i_1, ..., i_p$ and the sign is the one of the permutation $(j_1, .., j_r) \longmapsto (i_1, ..., i_p, s_1, ..., s_{r-p})$. Thus the linear map $\perp$ associated with e and e^* is an isomorphism. ∎

A *decomposable p-vector* of E is a p-vector of $\bigwedge^p E$ of type $X_1 \wedge ... \wedge X_p$. We have the similar notion for decomposable exterior form.

Proposition 1.7 *The linear map*

$$\perp: \bigwedge{}^n E \longrightarrow \bigwedge{}^{n-r} E^*$$

carries decomposable elements into decomposable elements.

Proof. A decomposable element of degree n goes into $\bigwedge^0 E$. In this case the proposition is obvious. Let $e_1 \wedge ... \wedge e_r$ be a decomposable element of degree r and we complete $(e_1, ..., e_r)$ to a basis $(e_1, ..., e_r, e_{r+1}, ..., e_n)$ such that, $e_1 \wedge ... \wedge e_n = e$. Then,

$$\perp (e_1 \wedge ... \wedge e_r) = i(e_1 \wedge ... \wedge e_r)e^* = \pm e^{r+1} \wedge ... \wedge e^n,$$

which is decomposable and corresponds to the space spanned by $n - r$ independent forms $e^{r+1}, ..., e^n$. ∎

1.3.2 Derivatives. Antiderivatives

Definition 1.6 *A linear endomorphism h of $\bigwedge E^*$ is a derivative of $\bigwedge E^*$ if it satisfies:*

1. *for every p-exterior form α on E, $h(\alpha)$ is an exterior form whose degree has the same parity as α;*

2. *$h(\alpha \wedge \beta) = h(\alpha) \wedge \beta + \alpha \wedge h(\beta)$ for every $\alpha, \beta \in \bigwedge E^*$.*

Definition 1.7 *A linear endomorphism h of $\bigwedge E^*$ is an anti-derivative of degree $q \in \mathbb{Z}$ of $\bigwedge E^*$ if it satisfies:*

1. *$h(\bigwedge^p E^*) \subseteq \bigwedge^{p+q} E^*$;*

2. $(-1)^{\deg(\alpha)} = (-1)^{\deg(h(\alpha))+1}$ *for every* α *of a given degree* $\deg(\alpha)$,

3. $h(\alpha \wedge \beta) = h(\alpha) \wedge \beta + (-1)^p \alpha \wedge h(\beta)$ *for every* $\alpha \in \bigwedge^p E^*$, $\beta \in \bigwedge E^*$.

The second condition implies that the degree of an anti-derivative necessarily is odd.

Example 1 *The interior product by a vector* $X \in E$ *is an antiderivative of degree* -1.

1.3.3   On the structure of the associative algebra $\bigwedge E^*$

Provided with the exterior product, $\bigwedge E^*$ is a non-commutative unitary associative algebra of dimension 2^n, where n is the dimension of E. Let us describe this structure for $\dim E = 1$ or 2.

1. $n = 1$. The algebra $\bigwedge E^*$ is of dimension 2. A basis is given by $\{1, e_1\}$ with $1 \in \bigwedge^0 E^* = \mathbb{R}$ and $e_1 \in E^* = \bigwedge^1 E^*$:

$$1 \wedge 1 = 1 \; , \; 1 \wedge e_1 = e_1 \wedge 1 = e_1 \; , \; e_1 \wedge e_1 = 0.$$

2. $n = 2$. If $\{e_1, e_2\}$ is a basis of E^* then the set $\{1, e_1, e_2, e_3 = e_1 \wedge e_2\}$ is a basis of $\bigwedge E^*$ and the exterior product is defined by:

$$1 \wedge e_i = e_i \wedge 1 = e_i \; , \quad i = 1, 2, 3$$

$$e_1 \wedge e_2 = -e_2 \wedge e_1 = e_3,$$

other products being zero.

The exterior algebra is a Clifford algebra associated with the null quadratic form. In fact, let q be a quadratic form on a vector space V. Let us consider the tensor algebra $T(V)$ on V. We define the ideal $I_q(V)$ as the ideal of $T(V)$ generated by the elements of the form $v \otimes v + q(v).1$ (1 is the identity of $T^0(V) = \mathbb{R}$). The Clifford algebra associated to q and V is the quotient algebra $T(V)/I_q(V)$. If $q = 0$ then $I_0(V)$ is generated by the vectors $v \otimes v$. The quotient is well isomorphic to $\bigwedge V$.

1.3.4   The graded ideals of $\bigwedge E^*$

An ideal I of $\bigwedge E^*$ is a subalgebra satisfying:

$$\alpha \wedge \omega \in I,$$

for all $\alpha \in I$ and $\omega \in \bigwedge E^*$.

Definition 1.8 *An ideal I of $\bigwedge E^*$ is called a graded ideal if it admits the following vectorial decomposition:*

$$I = I_0 \oplus I_1 \oplus \cdots \oplus I_n,$$

with $I_k = I \cap \bigwedge^k E^*$.

Example 2 *Let us consider the following subalgebra*

$$\bigwedge_i E = \bigwedge^i E \ \oplus \bigwedge^{i+1} E \oplus \cdots \oplus \bigwedge^n E.$$

It is a graded ideal of $\bigwedge E^*$. As this ideal has a non-zero square if $i < n/2$, the algebra $\bigwedge E^*$ is not simple.

Let I be a graded ideal of $\bigwedge E^*$. Every element $\alpha \in I$ admits a decomposition in homogeneous components:

$$\alpha = \alpha_0 + \alpha_1 + \cdots + \alpha_n, \quad \alpha_i \in \bigwedge^i E.$$

As I is graded, this decomposition is the decomposition associated with the graduation of I.

Lemma 3 *If $I_0 \neq \{0\}$ then $I = \bigwedge E^*$.*

Proof. In fact, $\bigwedge^0 E \cap I = I_0$. As I is an ideal, $I_0 \neq \{0\}$ implies $I_0 = \mathbb{R}$. Then, every linear form on E belongs to I. As these forms generate $\bigwedge E^*$ we then $\bigwedge E^* = I$. ∎

1.3.5   Generators of a graded ideal

Let us consider the subspace $A(I) \subset E$ defined by:

$$A(I) = \{X \in E \mid i(X)\alpha \in I , \forall \alpha \in I\}.$$

Its orthogonal space is, by definition, the subspace of $\bigwedge E^*$ given by:

$$A(I)^\perp = \{f \in E^* \mid f(X) = 0 , \forall X \in A(I)\}.$$

We denote by $\bigwedge A(I)^\perp$ the subalgebra of $\bigwedge E^*$ generated by $A(I)^\perp$.

Proposition 1.8 *Let I be a graded ideal of $\bigwedge E^*$ verifying $I_0 = \{0\}$. There exists a system of generators of I constituted by elements of $\bigwedge A(I)^\perp$.*

Proof. In fact, let $\{f_1, \cdots, f_r\}$ be a basis of $A(I)^{\perp}$. We complete this system in a basis $\{f_1, \cdots, f_r, h_{r+1}, \cdots, h_n\}$ of $\bigwedge^1 E^*$. Let us suppose that $I_1 \neq \{0\}$. Let $\{\alpha_1, \cdots, \alpha_k\}$ be a system of generators of I_1. The decomposition of these vectors on the fixed basis of $\bigwedge^1 E^*$ is written:

$$\alpha_i = \sum_{j=1}^{r} a_i^j f_j + \sum_{j=r+1}^{n} b_i^j h_j.$$

Consider a basis $\{Y_1, .., Y_r, X_1, \cdots, X_{n-r}\}$ of E of which the dual basis is $\{f_1, \cdots, f_r, h_{r+1}, \cdots, h_n\}$. Then, the system $\{X_1, \cdots, X_{n-r}\}$ forms a basis of $A(I)$. These vectors satisfy $f_i(X_j) = 0$ for all i and j. Thus

$$i(X_k)\alpha_i = \sum_{j=r+1}^{n} b_i^j h_j(X_k) \in I_0.$$

As, by hypothesis, $I_0 = \{0\}$, we have $i(X_k)\alpha_i = 0$. Moreover, we have $h_{r+j}(X_k) = \delta_{jk}$. This implies that $b_i^j = 0$ and

$$\alpha_i = \sum_{j=1}^{r} a_i^j f_j.$$

Thus, $\{f_1, \cdots, f_r\}$ generates I_1. We shall now prove that the exterior products $f_{i_1} \wedge \cdots \wedge f_{i_t} \wedge h_{i_{t+1}} \cdots \wedge h_{i_k}$ generate the component I_k. Every element $\beta \in I_k$ can be written:

$$\beta = \sum a_{i_1 i_2 \cdots i_k} f_{i_1} \wedge \cdots \wedge f_{i_k} + \Theta,$$

where Θ is a k-exterior form of the type

$$\Theta = \sum b_{i_1 i_2 \cdots i_k} g_{i_1} \wedge \cdots \wedge g_{i_k},$$

with, when $b_{i_1 i_2 \cdots i_k} \neq 0$, at least one of the g_{i_j} equal to h_{i_j}. Thus for every vector X_j of $A(I)$ we have:

$$i(X_j)\beta = \sum a_{i_1 i_2 \cdots i_k} i(X_j)(f_{i_1} \wedge \cdots \wedge f_{i_k}) + i(X_j)\Theta$$

and $i(X_j)\beta \in I_{k-1}$. But

$$i(X_j)(f_{i_1} \wedge \cdots \wedge f_{i_k}) = 0.$$

So $i(X_j)\Theta \in I_{k-1}$. If the exterior form Θ contains a term of the form $h_{i_1} \wedge \cdots \wedge h_{i_k}$, then $i(X_j)\Theta$ contains at least one term of the form $h_{i_1} \wedge$

$\cdots \wedge h_{i_{k-1}}$ and this term will be in I_{k-1}. By induction, we suppose that I_{k-1} is generated by $\bigwedge A(I)^{\perp}$. Then $h_{i_1} \wedge \cdots \wedge h_{i_{k-1}}$ cannot appear in the decomposition of Θ, and each term of this decomposition contains at least a form f_i. Then $\bigwedge A(I)^{\perp}$ generates the component I_k and the graded ideal I. Now suppose that $I_1 = \{0\}$. In this case we consider the smallest integer $i, i \geq 1$ such as $I_i \neq \{0\}$. The same arguments as above permit the completion of the proof. ∎

1.4   Linear system associated to a p-form

Let α be an exterior p-form on E. From a basis $\{e_1, e_2, \cdots, e_n\}$ of E we can define linear forms $\alpha_{i_1 i_2 \cdots i_{p-1}}$ on E by:

$$\alpha_{i_1 i_2 \cdots i_{p-1}}(X) = \alpha(X, e_{i_1}, e_{i_2}, \cdots, e_{i_{p-1}})$$

for all $X \in E$. This linear system

$$\{\alpha_{i_1 i_2 \cdots i_{p-1}} = 0 \ ,$$

$1 \leq i_1 < i_2 < \cdots < i_{p-1} \leq n$, obviously depends on α and on the basis $\{e_1, e_2, \cdots, e_n\}$. However, it enables one to construct a linear space which depends only on α and does not depend on the basis.

1.4.1   Linear system associated to a p-form

Definition 1.9 *The linear subspace $A^*(\alpha)$ of $\bigwedge^1 E = E^*$ generated by the linear form $\{\alpha_{i_1 i_2 \cdots i_{p-1}}\}$ is called the associated linear system of α.*

Of course, this linear space does not depend on the choice of the basis $\{e_1, e_2, \cdots, e_n\}$. In fact, let $\{e'_1, e'_2, \cdots, e'_n\}$ be a second basis of E. Let us denote by $(\alpha'_{i_1 i_2 \cdots i_{p-1}})$ the corresponding linear system. If $e'_i = \sum a_i^j e_j$ then

$$\begin{aligned}\alpha'_{i_1 i_2 \cdots i_{p-1}}(X) &= \alpha(X, e'_{i_1}, e'_{i_2}, \cdots, e'_{i_{p-1}}) \\ &= \sum a_{i_1}^{j_1} \cdots a_{i_{p-1}}^{j_{p-1}} \alpha_{j_1 j_2 \cdots j_{p-1}}(X).\end{aligned}$$

This proves that the linear space generated by the linear forms $\alpha_{i_1 i_2 \cdots i_{p-1}}$ coincides with the linear space generated by the forms $\alpha'_{i_1 i_2 \cdots i_{p-1}}$.

Definition 1.10 *The dimension of $A^*(\alpha)$ is called the rank of α and it is denoted by rank(α) (or $r(\alpha)$).*

The construction of $A^*(\alpha)$ shows that this linear space is the smaller linear subspace of E^* such that the exterior p-form α can be decomposed in a product of linear forms belonging to this linear subspace. So if $\{f_1, f_2, \cdots, f_r\}$ is a basis of $A^*(\alpha)$ then

$$\alpha = \sum_{1 \leq i_1 < \cdots < i_p \leq r} a^{i_1 i_2 \cdots i_p} f_{i_1} \wedge \cdots \wedge f_{i_p}.$$

We always have

$$\deg(\alpha) \leq \operatorname{rank}(\alpha),$$

$\deg(\alpha)$ being the degree of α. An important class of exterior forms corresponds to the forms having a degree equal to its rank. We will study these forms later.

1.4.2 The linear subspace associated to α

Let $A(\alpha)$ be the linear subspace of E defined by:

$$A(\alpha) = \{X \in E \mid i(X)\alpha = 0\}.$$

Theorem 1.1 *We have:*

$$A(\alpha) = (A^*(\alpha))^{\perp} = \{X \in E \mid f(X) = 0 \;\; \forall f \in A^*(\alpha)\}.$$

Proof. Let $\{f_1, f_2, \cdots, f_r\}$, $r = \operatorname{rank}(\alpha)$, be a basis of $A^*(\alpha)$. The exterior form α is written:

$$\alpha = \sum_{1 \leq i_1 < \cdots < i_p \leq r} a^{i_1 i_2 \cdots i_p} f_{i_1} \wedge \cdots \wedge f_{i_p}.$$

Let be $X \in (A^*(\alpha))^{\perp}$. Then $f_i(X) = 0$ for all i. Hence $i(X)\alpha = 0$ and $X \in A(\alpha)$. This gives $(A^*(\alpha))^{\perp} \subset A(\alpha)$. Conversely, let $X \in A(\alpha)$. Let us consider a basis $\{e_1, e_2, \cdots, e_n\}$ of E. The linear forms

$$\alpha_{i_1 i_2 \cdots i_{p-1}} = i(e_{i_1})i(e_{i_2})\cdots i(e_{i_{p-1}})\alpha$$

generate $A^*(\alpha)$. As $X \in E$,

$$i(X)\alpha = 0 \text{ and } i(X)\alpha_{i_1 i_2 \cdots i_{p-1}} = 0.$$

So, $f(X) = 0$ for all $f \in A^*(\alpha)$. This proves the theorem. ∎

Corollary 1.1 $\operatorname{rank}(\alpha) = \operatorname{co dim} A(\alpha)$

Example 4 *If α is a form of degree equal to 0, then $\text{rank}(\alpha) = 0$ and if α is a non zero form of degree 1 then $\text{rank}(\alpha) = 1$.*

Proposition 1.9 *If α is a non zero linear form on E, the associated subspace of α is the hyperplane of E defined by α.*

This is a direct consequence of the definition.

Remark 2 *The study of the space $A^*(\alpha)$ is interesting for the classification of the exterior p-forms. In fact, the dimension of these spaces is an invariant, up to isomorphism, of the form α. In the following section we shall see that this invariant gives the complete classification in the case where the form is of degree 2. For a form of any degree it is little more complicated. The classification necessitates the study of new invariants.*

1.5   Exterior 2-forms

1.5.1   Rank of an exterior 2-form

Let $\{e_1, e_2, ..., e_n\}$ be a basis of E and $\{f_1, f_2, ..., f_n\}$ the dual basis in $\bigwedge^1 E$. An exterior 2-form on E can be written:

$$\alpha = \alpha^{12} f_1 \wedge f_2 + \alpha^{13} f_1 \wedge f_3 + ... + \alpha^{n-1n} f_{n-1} \wedge f_n.$$

The associated linear system is generated by the linear forms $i(e_j)\alpha$. This system is given by:

$$\begin{cases} \alpha_1 = -\alpha^{12} f_2 - \alpha^{13} f_3 - ... - \alpha^{1n} f_n = 0, \\ \alpha_2 = +\alpha^{12} f_1 - \alpha^{23} f_3 - ... - \alpha^{2n} f_n = 0, \\ \qquad\qquad \vdots \\ \alpha_n = \alpha^{1n} f_1 + \alpha^{2n} f_2 + ... + \alpha^{n-1n} f_{n-1} = 0 \ . \end{cases}$$

Its rank is equal to the rank of the antisymmetric matrix

$$\begin{pmatrix} 0 & \alpha^{12} & \alpha^{13} & \cdots & \alpha^{1n} \\ -\alpha^{12} & 0 & \alpha^{23} & \cdots & \alpha^{2n} \\ \cdots & \cdots & \cdots & \cdots & \cdots \\ \cdots & \cdots & \cdots & \cdots & \cdots \\ -\alpha^{1n} & -\alpha^{2n} & -\alpha^{3n} & \cdots & 0 \end{pmatrix}.$$

Proposition 1.10 *Every exterior 2-form on a vector space E is of even rank.*

Let us consider the exterior 2-form α. Let $X = \sum X_i e_i$ and $Y = \sum Y_i e_i$ be two vectors of E. Thus

$$\alpha(X,Y) = \sum_{i<j} \alpha^{ij}(X_i Y_j - X_j Y_i).$$

The partial derivative of the form α, with respect to the variable X_i, is defined by:

$$\frac{\partial \alpha}{\partial X_i}(X) = \sum_{i<j} \alpha^{ij} X_j.$$

Then we have:

$$2\alpha = \sum \frac{\partial \alpha}{\partial X_i} \wedge f_i,$$

and the associated linear system of α is generated by the linear forms

$$\frac{\partial \alpha}{\partial X_1}, \ldots, \frac{\partial \alpha}{\partial X_n}.$$

This corresponds to the determination of the rank of a symmetric bilinear form:

$$\begin{aligned}
F(X,Y) &= \sum a_{ij} X_i Y_j \\
&= \sum X_i \frac{\partial F}{\partial Y_i} \\
&= \sum Y_i \frac{\partial F}{\partial X_i}.
\end{aligned}$$

The reduction of the linear system

$$\begin{cases} \frac{\partial F}{\partial X_1} = 0, \\ \ldots \\ \frac{\partial F}{\partial X_n} = 0 \end{cases}$$

is equivalent to that of the symmetric form F.

1.5.2 Reduction of an exterior 2-form

Lemma 5 *There is a basis* $\{v_1, v_2, \ldots, v_n\}$ *of* E *such that:*

1. $\alpha(v_1, v_2) = \alpha(v_3, v_4) = \ldots = \alpha(v_{2p-1}, v_{2p}) = 1$,

2. $\alpha(v_i, v_j) = 0$, $(i,j) \notin \{(1,2), \ldots, (2p-1, 2p), (2,1), \ldots, (2p, 2p-1)\}$.

Proof. Suppose that the form α satisfies $\alpha \neq 0$. Thus there are independent vectors v_1 and v_2 such as $\alpha(v_1, v_2) \neq 0$. Let us consider

$$H_i = \{x \in E \mid \alpha(v_i, x) = 0\},$$

$i = 1, 2$, and $H = H_1 \cap H_2$. If F is the linear space generated by the vectors v_1 and v_2, then $E = H \oplus F$, $\dim H = n - 2$. The restriction of the exterior form α to the linear space H is of rank $2p - 2$. By induction on the dimension of the space E we complete the proof. ∎

Theorem 1.2 *For every exterior 2-form β of rank $2p$, on a vector space E, there exists a basis $\{f^1, f^2, \ldots, f^{2p}\}$ of E^* such that*

$$\beta = f^1 \wedge f^2 + f^3 \wedge f^4 + \ldots + f^{2p-1} \wedge f^{2p}.$$

Proof. The proof is based on the preceding lemma. Let us consider the basis $\{v_1, \ldots, v_n\}$ given by the previous lemma. If $\{f^1, f^2, \ldots, f^n\}$ is the dual basis, the exterior 2-form

$$\beta = f^1 \wedge f^2 + f^3 \wedge f^4 + \ldots + f^{2p-1} \wedge f^{2p}$$

satisfies $(\beta - \alpha)(v_{2i-1}, v_{2i}) = 0$ for $i = 1, 2, \ldots, p$. As it also satisfies $(\beta - \alpha)(v_i, v_j) = 0$, then $\beta - \alpha = 0$. ∎

1.5.3   Determination of the rank

Proposition 1.11 *Let α be a non-trivial exterior 2-form on E. Then the rank of α is $2p$ if and only if we have*

$$\alpha^p \neq 0, \quad \alpha^{p+1} = 0.$$

In fact, we can write

$$\alpha = f^1 \wedge f^2 + f^3 \wedge f^4 + \ldots + f^{2p-1} \wedge f^{2p}.$$

1.6   Cartan's Lemma

Theorem 1.3 *Let $f_1, f_2, \ldots, f_p$ be independent linear forms on E. Let us consider linear forms $\phi_1, \phi_2, \ldots, \phi_p$ on E satisfying*

$$f_1 \wedge \phi_1 + f_2 \wedge \phi_2 + \ldots + f_p \wedge \phi_p = 0.$$

Then the linear forms ϕ_i verify

$$\phi_i = \sum_j a_{ij} f_i \wedge f_j$$

with $a_{ij} = a_{ji}$.

Proof. First suppose $p = n$. In this case one has $\phi_i = \sum a_{ij} f_j$, because, $\{f_1, f_2, \cdots, f_n\}$ is a basis of E^*. The equation

$$\sum \phi_i \wedge f_i = 0$$

implies

$$\sum a_{ij} f_j \wedge f_i = 0.$$

Since the linear forms f_j are independent this exterior 2-form is zero whenever its coefficients with respect to the basis $\{f_i \wedge f_j\}_{i<j}$ are zero. Thus $a_{ij} - a_{ji} = 0$, and the matrix (a_{ij}) is symmetric.

Now, suppose that $p < n$ and consider a basis $\{f_1, ..., f_p, g_{p+1}, ..., g_n\}$ of $\wedge^1 E^*$. We have:

$$\phi_i = \sum_{j=1}^{p} a_{ij} f_j + \sum_{j=1}^{n-p} a_{ij+p} g_{j+p} \cdot$$

The equation $\sum_{j=1}^{p} \phi_i \wedge f_i = 0$ implies

$$\sum_{i=1}^{p} \sum_{j=1}^{p} a_{ij} f_j \wedge f_i + \sum_{i=1}^{p} \sum_{j=1}^{n-p} a_{ij+p} g_{j+p} \wedge f_i = 0 .$$

We deduce immediately that $a_{ij+p} = 0$. This proves the theorem. ∎

1.7   Monomial Forms

Definition 1.11 *A non-trivial exterior p-form α on E is called a monomial p-form, if there exist p linearly independent linear forms $f_1, ..., f_p$ such as*

$$\alpha = f_1 \wedge ... \wedge f_p.$$

It is clear that if α is a monomial p-form then

$$\text{rank}(\alpha) = p = \deg(\alpha).$$

Theorem 1.4 *Let α be an exterior p-form on E. Then α is a monomial form if and only if*

$$\text{rank}\,(\alpha) = p = \deg(\alpha).$$

Proof. Let α be a p-form verifying $\text{rank}\,(\alpha) = p$. Let us consider a basis $\{f_1, ..., f_p\}$ of the linear space $A^*\,(\alpha)$ associated to α. According to the definition of this space one has

$$\alpha = a f_1 \wedge ... \wedge f_p,$$

and α is a monomial form. ■

Corollary 1.2 *A non-trivial exterior n-form on E, with $\dim E = n$, is of rank n.*

Proposition 1.12 *Every non-trivial exterior form of degree $n-1$, is a monomial form.*

Proof. Let α be an exterior form of degree $n-1$ on E. Consider the linear map

$$h : E^* \to \Lambda^n E$$

defined by $h(f) = f \wedge \alpha$. As $\dim \Lambda^n E = 1$, $\dim \ker h = n$ or $n-1$. We can find a basis $\{f^1, ..., f^n\}$ of E^* in order that $\{f^1, ..., f^{n-1}\}$ be a basis of $\ker h$. We can write

$$\alpha = \sum_{i=1}^{n} a_i f^1 \wedge ... f^{i-1} \wedge f^{i+1} \wedge ... \wedge f^n.$$

The map h is determined by

$$h(f^i) = (-1)^{i-1} a_i f^1 \wedge ... \wedge f^n.$$

Thus α is zero if $h = 0$. If $h \neq 0$ we have $a_i = 0$ for $i = 1, ..., n-1$, and α is a monomial form. ■

Corollary 1.3 *A non-trivial exterior form on E of degree $n-1$, is of rank $n-1$.*

Corollary 1.4 *A non-trivial exterior form on E, of degree $n-2$, is of rank $n-2$ or n.*

Proof. If $\deg(\alpha) = n - 2$, then $\text{rank}(\alpha) = n$, or $n - 1$ or $n - 2$. Suppose that $\text{rank}(\alpha) = n - 1$. Let $A(\alpha)$ be the associated linear space of α. The dimension of the quotient space $E/A(\alpha)$ is $n - 1$. As α is a form of degree $n - 2$ in $A(E/A(\alpha)) \subset A(E)$, α is monomial of rank $n - 2$. This is impossible. Therefore $\text{rank}(\alpha) \neq n - 1$.

Now consider the case where $\dim E = 4$. The exterior 2-form α defined by

$$\alpha = f^1 \wedge f^2 + f^3 \wedge f^4$$

is of rank $4 = n$. ∎

Chapter 2

EXTERIOR SYSTEMS

2.1   Exterior Systems

2.1.1   Systems of exterior equations

Let E be an n-dimensional real vector space and $\bigwedge E^* = \oplus \bigwedge^i E^*$ its exterior algebra. An exterior equation is an equation of the form

$$\theta = 0$$

where $\theta \in \bigwedge^p E^*$. A solution of this exterior equation is a linear subspace H of E verifying

$$\theta(X_1, X_2, \cdots, X_p) = 0 \ ,$$

for all $X_1, X_2, \cdots, X_p \in H$.

In the case $p = 1$ the corresponding exterior equation is a linear equation. For $p = 2$ the exterior form θ is an alternating bilinear form, and every linear subspace contained in the kernel of this bilinear form is a solution.

Definition 2.1 *An exterior system is defined by a finite number of exterior equations on E:*

$$\begin{cases} \theta_1 = 0, \\ \theta_2 = 0 \\ \quad \vdots \\ \theta_r = 0 \end{cases}$$

with $\theta_i \in \bigwedge^{p_i} E^*$.

Later we will study the important particular case in which each form θ_i is of degree 2.

2.1.2 Solution of an exterior system

Definition 2.2 *A solution of the exterior system* $\{\theta_1 = 0, ..., \theta_r = 0\}$ *is a linear subspace* H *of* E *verifying* $\theta_i(H) = 0$, $i = 1, ..., r$, *that is,*

$$\theta_i\left(X_{i_{s_1}}, ..., X_{i_{s_i}}\right) = 0$$

for all $i = 1, \cdots, r$, *with* $s_i = \deg(\theta_i)$ *and* $X_{i_{s_j}} \in H$.

2.1.3 Examples

1. Suppose that the equations of the exterior system are given by linear forms. Then every subspace contained in the kernel of the exterior system is a solution.

2. Let us consider the system defined on $\mathbb{R}^{2n}$ by the equation

$$\theta = f_1 \wedge f_2 + f_3 \wedge f_4 + \cdots + f_{2n-1} \wedge f_{2n},$$

   where $\{f_1, ..., f_{2n}\}$ is the dual basis of a given basis $\{e_1, ..., e_{2n}\}$ of $\mathbb{R}^{2n}$. Then the subspace H generated by the vectors $\{e_1, e_3, ..., e_{2n-1}\}$ is a solution of the exterior system $\theta = 0$. It is clear that this solution is maximal (for the inclusion relation). It is called a Lagragian subspace of $(\mathbb{R}^{2n}, \theta)$. Of course, such a maximal solution is not unique.

2.2 Resolution of the equation $\theta = 0$ with $\theta \in \Lambda^q(E)$

In this section, we study the particular case of an exterior system with only one exterior equation of degree q

$$\theta = 0.$$

We will determine the linear subspace H of dimension p satisfying:

$$\theta(H) = 0,$$

that is,

$$\theta(X_1, ..., X_q) = 0$$

for all $X_1, ..., X_q \in H$.

2.2.1   Plücker coordinates

Let H be a p-plane of $\mathbb{R}^n$, let $\{e_1, ..., e_n\}$ be a basis of $\mathbb{R}^n$ and let $\{v_1, ..., v_p\}$ be a basis of the subspace H. Let us put

$$v_i = \sum_{j=1}^{n} \alpha_i^j e_j \quad (i = 1, ..., p).$$

The matrix $M = \left(\alpha_i^j \right)$ is of rank p. Let us denote by

$$u^{i_1 ... i_p} \quad (i_1 < i_2 < ... < i_p),$$

the determinant of the matrix of order p constituted by the columns $i_1, i_2, ...,$ i_p of the matrix M:

$$u^{i_1 ... i_p} = \det \begin{pmatrix} \alpha_1^{i_1} & ... & \alpha_1^{i_p} \\ & \vdots & \\ \alpha_p^{i_1} & ... & \alpha_p^{i_p} \end{pmatrix}.$$

Definition 2.3 *The $\{u^{i_1 ... i_p}\}$ with $1 \le i_1 < i_2 < ... < i_p \le n$ are called the Plücker coordinates of the linear subspace H.*

As the dimension of H is p, there exists $i_1, i_2, ..., i_p$ such that $u^{i_1 ... i_p} \neq 0$. Therefore we can consider the Plücker coordinates as a system of coordinates of the projective kind.

2.2.2   Examples

1. Let H be a one-dimensional subspace of $\mathbb{R}^3$. It is generated by a non-zero vector $v = (a, b, c)$. The Plücker coordinates are the components (a, b, c) of this vector. It is clear that these coordinates entirely define this vector subspace.

2. Let H be a 2-dimensional vector subspace of $\mathbb{R}^3$. Let $\{ u = (a, b, c)$, $v = (a', b', c')\}$ be a basis of H. The rank of the matrix

$$M = \begin{pmatrix} a & b & c \\ a' & b' & c' \end{pmatrix}$$

is 2. There is then a non-trivial minor of order 2. Let us note (u^{ij}) these 2-minors:

$$u^{12} = ab' - ba' \ , \ u^{13} = ac' - ca' \ , \ u^{23} = bc' - cb'.$$

The Plücker coordinates of H are the parameters (u^{12}, u^{13}, u^{23}). They are the components of a directional vector of this plane.

3. The Plücker coordinates of a 2-plane H in $\mathbb{R}^4$ are defined as follows : let $\{e_1, e_2, e_3, e_4\}$ be a basis of $\mathbb{R}^4$ and let $\{v_1, v_2\}$ be a basis of H. Let us consider the matrix

$$M = \begin{pmatrix} \alpha_1^1 & \alpha_2^1 & \alpha_3^1 & \alpha_4^1 \\ \alpha_1^2 & \alpha_2^2 & \alpha_3^2 & \alpha_4^2 \end{pmatrix}$$

of the components of the vectors $v_i \in H$ with respect to the basis $\{e_i\}$. The minors

$$u^{12} = \alpha_1^1 \alpha_2^2 - \alpha_1^2 \alpha_2^1, \quad u^{13} = \alpha_1^1 \alpha_2^3 - \alpha_2^1 \alpha_1^3, \quad u^{14} = \alpha_1^1 \alpha_2^4 - \alpha_1^4 \alpha_2^1,$$
$$u^{23} = \alpha_1^2 \alpha_2^3 - \alpha_1^3 \alpha_2^2, \quad u^{24} = \alpha_1^2 \alpha_2^4 - \alpha_1^4 \alpha_2^2, \quad u^{34} = \alpha_1^3 \alpha_2^4 - \alpha_1^4 \alpha_2^3$$

define a system of Plücker coordinates of H.

2.2.3   Change of Plücker coordinates

Let H be a p-plane, that is a p-dimensional vectorial subspace in $\mathbb{R}^n$. Let us fix a basis $\{e_1, ..., e_n\}$ of $\mathbb{R}^n$. The Plücker coordinates of H are defined from an arbitrary basis $\{v_1, ..., v_p\}$ of H, and thus depend on this choice. We will examine the effect of a change of bases on the writing of these Plücker coordinates.

Let $\{w_1, ..., w_p\}$ be a new basis of H and let us denote by P the matrix of the change of basis. Let $M = \left(\alpha_i^j\right)$ and $M' = \left(\beta_i^j\right)$ be the matrices defined by

$$v_i = \sum_{j=1}^n \alpha_i^j e_j,$$

and

$$w_i = \sum_{j=1}^n \beta_i^j e_j.$$

These matrices are connected by the following relation:

$$M' = MP.$$

Let us denote by $u^{i_1 \cdots i_p}$ the Plücker coordinates of H, with respect to the basis $\{v_1, ..., v_p\}$, and $u'^{i_1 \cdots i_p}$ the Plücker coordinates of H with respect to the basis $\{w_1, ..., w_p\}$. The above relationship implies:

$$u'^{i_1 \cdots i_p} = \lambda u^{i_1 \cdots i_p}$$

with $\lambda = \det P \neq 0$.

These formulae show well that the change of Plücker coordinates corresponds to a change of projective coordinates.

2.2.4 Determination of a p-plane

Let H be a p-plane in $\mathbb{R}^n$, let $\{e_1, ..., e_n\}$ be a fixed basis of $\mathbb{R}^n$ and $\{v_1, ..., v_p\}$ a basis of H. Let $\left(u^{i_1 \cdots i_p} \right)$ be the Plücker coordinates of H with respect to this basis. Let

$$v_i = \sum_{j=1}^{n} \alpha_i^j e_j \quad (i = 1, ..., p).$$

A system of equations of H is formed by writing that the matrix

$$\begin{pmatrix} x_1 & \cdots & x_n \\ \alpha_1^1 & \cdots & \alpha_1^n \\ \vdots & & \\ \alpha_p^1 & \cdots & \alpha_p^n \end{pmatrix}$$

is of rank p, that is, by cancelling all the minors of order $p+1$. By expanding with respect to the first line, each of these minors of order $p+1$, one finds linear equations in the variables x_i whose coefficients are the minors of order p of the matrix $\left(\alpha_j^i \right)$. However, these last minors are the Plücker coordinates of H. Thus one finds a system of linear equations defining H whose coefficients are the Plücker coordinates of H.

Example 6 *For a 2-plane in $\mathbb{R}^3$ one has:*

$$\det \begin{pmatrix} x & y & z \\ a & b & c \\ a' & b' & c' \end{pmatrix} = 0.$$

Thus the equation

$$x \left(bc' - cb' \right) - y \left(ac' - ca' \right) + z \left(ab' - ba' \right) = 0,$$

that is,

$$xu^{23} - yu^{13} + zu^{12} = 0.$$

We can note that any system (α, β, γ) is a system of Plücker coordinates of a 2-plane of $\mathbb{R}^3$ if one of its parameters is non-zero. In the general case, the situation is more delicate; in fact, a system $\left(u^{i_1 \cdots i_p} \right)$ with $1 \leq i_1 < i_2 < \ldots < i_p \leq n$ does not necessarily define a system of Plücker coordinates of a p-plane in $\mathbb{R}^n$. In the following subsection, we will describe the necessary and sufficient conditions in order that, such a system corresponds to Plücker coordinates of a p-plane.

2.2.5   Characterization of Plücker coordinates

Let $\{u^{i_1 \cdots i_p}\}$ be a system of real numbers with $1 \leq i_1 < i_2 < \ldots < i_p \leq n$; we suppose that there exist $i_1, i_2, ..., i_p$ so that $u^{i_1 \cdots i_p} \neq 0$. The aim of this section is to describe the conditions on the system $\left(u^{i_1 \cdots i_p} \right)$ for this system to represent the Plücker coordinates of a p-plane H. Following the definition, a necessary condition is

$$u^{i_1 \cdots i_j i_{j+1} \cdots i_p} = -u^{i_1 \cdots i_{j+1} i_j \cdots i_p}.$$

Assume that this condition is satisfied and let us consider the exterior p-form θ given by :

$$\theta = \sum u^{i_1 \cdots i_p} f^{i_1} \wedge \cdots \wedge f^{i_p},$$

where $\{f^1, \cdots, f^n\}$ is the dual basis of a given basis $\{e_1, ..., e_n\}$ of $\mathbb{R}^n$.

Theorem 2.1 *The system $\{u^{i_1 \cdots i_p}\}$ is a system of Plücker coordinates of a p-plane H if and only if the exterior p-form θ is a monomial form.*

Proof. Let us suppose that there is a p-plane H whose Plücker coordinates are $u^{i_1 \cdots i_p}$. Let us consider the basis $\{\widetilde{e}_1, \cdots, \widetilde{e}_p\}$ of H associated with these coordinates and note $\{g_1, \cdots, g_p\}$ the dual basis. The definition of these coordinates shows that:

$$\theta = g_1 \wedge \cdots \wedge g_p.$$

Thus it is a monomial exterior form. Conversely, if θ is a monomial exterior form, that is, $\theta = h_1 \wedge ... \wedge h_p$, then the plane H whose basis is given by the dual vectors $\{v_1, ..., v_p\}$ of the forms $\{h_1, ..., h_p\}$ has as Plücker coordinates $\left(u^{i_1 \cdots i_p} \right)$. ■

Example 7 *Study of the Grassmann manifold $G_{2,4}$.*

Let us consider the system $\left(u^{12}, u^{13}, u^{14}, u^{23}, u^{24}, u^{34}\right)$, at least one of the coordinates u^{ij} being non-zero and set $u^{ij} = -u^{ji}$ if $i \geq j$. Let us seek the conditions for this system to be a system of Plücker coordinates of a 2-plane in $\mathbb{R}^4$. Let us consider the exterior 2-form:

$$\theta = u^{12} f_1 \wedge f_2 + u^{13} f_1 \wedge f_3 + u^{14} f_1 \wedge f_4 + u^{23} f_2 \wedge f_3 + u^{24} f_2 \wedge f_4 + u^{34} f_3 \wedge f_4.$$

Then θ is a monomial exterior form if and only if it is of rank 2, that is, if and only if its associated linear system:

$$\begin{cases} \alpha_1 = u^{12} f_2 + u^{13} f_3 + u^{14} f_4, \\ \alpha_2 = -u^{12} f_1 + u^{23} f_3 + u^{24} f_4, \\ \alpha_3 = -u^{13} f_1 - u^{23} f_3 + u^{34} f_4, \\ \alpha_4 = -u^{14} f_1 - u^{24} f_2 - u^{34} f_3 \end{cases}$$

is of rank 2. This is equivalent to saying that the matrix

$$A = \begin{pmatrix} 0 & u^{12} & u^{13} & u^{14} \\ -u^{12} & 0 & u^{23} & u^{24} \\ -u^{13} & -u^{23} & 0 & u^{34} \\ -u^{14} & -u^{24} & -u^{34} & 0 \end{pmatrix}$$

is of rank 2. As this matrix is antisymmetric, this condition is equivalent to the condition:

$$\det(A) = 0$$

(recall that $A \neq 0$). Then we have proved :

Proposition 2.1 *The system* $(u^{ij}), 1 \leq i \leq j \leq 4$, *defines a system of Plücker coordinates of a 2-plane in $\mathbb{R}^4$ if and only if*

$$u^{12} u^{34} - u^{13} u^{24} + u^{23} u^{14} = 0.$$

Thus this last equation represents the equation of the Grassmann manifold

$$G_{2,4} = \left\{ \text{2-planes in } \mathbb{R}^4 \right\},$$

which appears as an homogeneous algebraic submanifold of the projective space $\mathcal{P}\left(\mathbb{R}^6\right)$.

2.2.6   Study of the equation $\theta = 0$ with $\theta \in \bigwedge^q E$

It is assumed that θ is an exterior q-form in the vector space E. Let $\{e_1, \cdots, e_n\}$ be a fixed basis of E and let $\{f_1, \cdots, f_n\}$ be its dual basis. The exterior form θ is written :

$$\theta = \sum \theta_{j_1 \cdots j_q} f_{j_1} \wedge \cdots \wedge f_{j_q}.$$

The q-planes H solutions of $\theta = 0$

Let H be a q-plane in E and let $\{h_1, \ldots, h_q\}$ be a basis of H; let us denote by $\left(u^{i_1 \cdots i_q}\right)$ the corresponding Plücker coordinates of H, the basis of E being fixed. If H is a solution of $\theta = 0$ then

$$
\begin{aligned}
\theta\left(h_1, \cdots, h_q\right) &= \sum \theta_{j_1 \cdots j_q} f_{j_1} \wedge \ldots \wedge f_{j_q}\left(h_1, \cdots, h_q\right) \\
&= \sum \theta_{j_1 \cdots j_q} u^{j_1 \cdots j_q} \\
&= 0
\end{aligned}
$$

Thus H is a solution if and only if its Plücker coordinates verify:

$$\sum \theta_{j_1 \cdots j_q} u^{j_1 \cdots j_q} = 0.$$

This condition is completely independent of the chosen basis in H.

Proposition 2.2 *A q-plane $H \subset E$ with Plücker coordinates $\left(u^{i_1 \cdots i_q}\right)$ is a solution of the exterior equation*

$$\theta = \sum \theta_{j_1 \cdots j_q} f_{j_1} \wedge \ldots \wedge f_{j_q} = 0,$$

if and only if we have

$$\sum \theta_{j_1 \cdots j_q} u^{j_1 \cdots j_q} = 0.$$

The p-planes solution with $p > q$

Proposition 2.3 *A p-plane H is a solution of the exterior equation*

$$\theta = 0, \quad \theta \in \Lambda^q E$$

with $p > q$, if and only if, every q-plane contained in H is solution of this equation.

Proof. It is clear that if H is a solution and if H_1 is contained in H, then $\theta(H_1) = 0$.

Conversely, let us suppose that every q-plane contained in H is a solution. Let us suppose that H is not a solution. Then, there are q independent vectors $(X_1, \cdots, X_q)$ in H satisfying:

$$\theta(X_1, \cdots, X_q) \neq 0.$$

The subspace H_1 of H generated by $X_1, \cdots, X_q$ is of dimension q. It is not a solution of $\theta = 0$. This gives a contradiction and H is solution to $\theta = 0$.

The p-planes solution with $p < q$

Proposition 2.4 *Every p-plane, with $p < q$, is a solution of the exterior equation*

$$\theta = 0, \quad \theta \in \Lambda^q E.$$

In fact, let $X_1, \ldots, X_q$ be q vectors in a given p-plane H. If $p < q$ these vectors are not independent and

$$\theta(X_1, \cdots, X_q) = 0.$$

2.3 Algebraically equivalent systems

2.3.1 Ideal generated by an exterior system

Let $(S) = (\theta_1 = 0, \cdots, \theta_r = 0)$ be an exterior system with $\deg(\theta_i) = p_i > 0$. Our purpose is to compare this system with other systems having the same solutions. Let

$$I(S) = \left\{ \alpha \in \bigwedge E^* \mid \alpha = \sum_{i=1}^{r} \phi_i \wedge \theta_i, \text{ with } \phi_i \in \bigwedge E^* \right\}.$$

By construction $I(S)$ is an ideal of $\bigwedge E^*$, called the ideal generated by the system (S). It is clear that any solution of the system (S) is a solution of the exterior equation $\alpha = 0$, whatever the form α belonging to $I(S)$.

Definition 2.4 *The system (S) is complete if any exterior form annihilating each solution of the system belongs to the ideal $I(S)$ generated by (S).*

2.3.2   Examples

1. Every linear system is complete.

2. Let us consider the exterior system in $\mathbb{R}^4$ given by:

$$\begin{cases} \theta_1 = f_1 \wedge f_3 = 0, \\ \theta_2 = f_1 \wedge f_4 = 0, \\ \theta_3 = f_1 \wedge f_2 - f_3 \wedge f_4 = 0. \end{cases}$$

Let H be a 2-plane in $\mathbb{R}^4$ with Plücker coordinates $\left(u^{ij}\right)$. These coordinates verify,

$$u^{12}u^{34} - u^{13}u^{24} + u^{14}u^{23} = 0.$$

As H is a solution of the exterior system we have

$$\theta_1\left(H\right) = \theta_2\left(H\right) = \theta_3\left(H\right) = 0,$$

which implies

$$\begin{cases} u^{13} = 0, \\ u^{14} = 0, \\ u^{12} - u^{34} = 0. \end{cases}$$

This linear system is thus reduced to:

$$\begin{cases} u^{13} = 0, \\ u^{14} = 0, \\ u^{12} = 0, \\ u^{34} = 0, \end{cases}$$

and any solution H of the exterior system has as Plücker coordinates

$$\left(0, 0, 0, u^{23}, u^{24}, 0\right).$$

Let us consider the exterior form $\beta = f_1 \wedge f_2$. It satisfies $\beta\left(H\right) = 0$ too because $u^{12} = 0$. This form is not in the ideal $I(S)$ and the exterior system (S) is not complete.

3. The exterior system in $\mathbb{R}^4$ given by

$$\begin{cases} \theta_1 = f_1 \wedge f_2 = 0, \\ \theta_2 = f_1 \wedge f_3 = 0 \end{cases}$$

is not complete.

2.3.3 Algebraically equivalent systems

Definition 2.5 *Two exterior systems* (S) *and* (S') *are algebraically equivalent if they generate the same ideal, that is, if* $I(S) = I(S')$.

Let us consider the exterior system in $\mathbb{R}^n$

$$\begin{cases} f_1 = 0, \\ f_2 = 0, \\ \theta = 0, \end{cases}$$

where f_1 and f_2 are linear forms and θ an exterior 2-form. Every algebraically equivalent system has the following form:

$$\begin{cases} af_1 + bf_2 = 0, \\ cf_1 + df_2 = 0, \\ e\theta + f_1 \wedge g + f_2 \wedge h = 0, \end{cases}$$

where $a, b, c, d, e \in \mathbb{R}$, $ad - bc \neq 0$, $e \neq 0$, and $g, h \in \Lambda^1 \mathbb{R}^n$.
It is clear that two algebraically equivalent systems have the same solutions. On the other hand two systems can have the same solutions without being algebraically equivalent.
The two exterior systems in $\mathbb{R}^4$

$$\begin{cases} f_1 \wedge f_3 = 0, \\ f_1 \wedge f_4 = 0, \\ f_1 \wedge f_2 - f_3 \wedge f_4 = 0, \end{cases} \quad \text{and} \quad \begin{cases} f_1 \wedge f_2 = 0, \\ f_1 \wedge f_3 = 0, \\ f_1 \wedge f_4 = 0, \\ f_3 \wedge f_4 = 0, \end{cases}$$

have the same solutions but are not algebraically equivalent. Indeed, the Plücker coordinates of the 2-plane solution of the first system satisfies $u^{13} = 0$, $u^{14} = 0$, $u^{12} = 0$, and $u^{34} = 0$. Such a plane is also a solution of the second system. These two systems are not algebraically equivalent because the form $f_3 \wedge f_4$ is not in the ideal generated by the first system.

Proposition 2.5 *If the system* (S) *is complete, every system* (S') *having the same solution as* (S) *is algebraically equivalent to* (S).

2.4 Vector space associated with an exterior system

Let (S) be an exterior system on E.

Definition 2.6 *The vector space associated with* (S) *is the smallest subspace* Q *of* E^* *such that* ΛQ *contains a subset generating* $I(S)$.

One deduces from this that the linear system

$$\left\{ \alpha = 0 , \quad \alpha \in \overset{1}{\bigwedge} Q \right.$$

is algebraically equivalent to (S).

Definition 2.7 *The rank of the exterior system* (S) *is the dimension of the associated linear space* Q.

Proposition 2.6 *Two algebraically equivalent systems have the same rank.*

This arises directly from the definition of the rank.
In the following chapter we will study a particular class of exterior systems in which each equation is of degree two. These systems correspond to the linear models of the equations of statistical mechanics introduced by Nambu.

Chapter 3

k-SYMPLECTIC
EXTERIOR SYSTEMS

Introduction.

The classical mechanics has led to the study of the exterior equation

$$\theta = 0,$$

where θ is an exterior 2-form of maximum rank. In the first chapter we have recalled the classification of exterior 2-forms, without returning to the classic study of vector spaces equipped with such forms.

We propose, in this chapter, to generalize this symplectic geometry by replacing the symplectic equation by an exterior system of symplectic type. We would have been able to widen the symplectic geometry by considering forms of high degree. We note that recent works envisage structures defined by an exterior equation of degree 3 (see [G.M] and [KI]).

One of our motivations for considering an exterior system of degree 2 rather than an exterior equation of degree 3 comes from mechanics. Doing so seems a good approach to formalizing the statistical mechanics pointed out by Nambu. Other motivations are related to the classification of the systems of bilinear alternating forms. The multi-symplectic systems we are going to study in this chapter appear naturally in the 3-dimensional case. In high dimension an infinity of systems which are not algebraically equivalent can be defined. The k-symplectic systems are defined directly by conditions of regularity; they can be interpreted as models of exterior systems of maximum rank.

3.1   Classification of exterior systems

Let (S) be an exterior system of $\mathbb{R}^n$ spanned by

$$\theta^1 = 0, \ldots, \theta^k = 0,$$

where $\theta^i = 0$ are exterior two forms $(\theta^i \in \Lambda^2(\mathbb{R}^n))$, and we write $(S) = \{\theta^1, ..., \theta^k\}$. Suppose that the two forms θ^i are independent in the real space $\Lambda^2(\mathbb{R}^n)^*$.

If $k = 1$ the classification of systems $(S) = \{\theta^1\}$ is determined by the rank of θ^1. In this section we propose to study the classification problem of such systems for $k \geq 2$.

It will be instructive to recall some fundamental definitions.

3.1.1   Rank of (S)

The associated space to the system (S) is the smallest subspace $A^*(S)$ of $(\mathbb{R}^n)^*$ such that $\bigwedge A^*(S)$ contains a subset spanning $I(S)$.

Definition 3.1 *The rank of the exterior system (S) is the dimension of the associated space $A^*(S)$.*

Notice that $A(S) = \{X \in \mathbb{R}^n \mid i(X)\theta = 0, \forall \theta \in S\}$. It is clear that the rank of (S) coincides with the codimension of $A(S)$.

3.1.2   Classification for $k = 2$ and $n = 3$

The case rank$(S) = 1$

In this case we have $\dim(A(S)) = 2$. Let $\{e_1, e_2\}$ be a basis of $A(S)$ which we extend to a basis $\{e_1, e_2, e_3\}$ of $\mathbb{R}^3$. Relative to the dual basis $\{\omega^1, \omega^2, \omega^3\}$ of $\{e_1, e_2, e_3\}$ the forms θ^1 and θ^2 can be written

$$\begin{cases} \theta^1 = \sum C^1_{ij}\omega^i \wedge \omega^j, \\ \theta^2 = \sum C^2_{ij}\omega^i \wedge \omega^j, \end{cases}$$

where C^l_{ij} are real constants. We have

$$i(e_1)\theta^l = \sum_j C^l_{1j}\omega^j = 0$$

and

$$i(e_2)\theta^l = \sum_j C^l_{2j}\omega^j = 0.$$

Thus $C^l_{jk} = 0$ for all $l, j = 1, 2$. This implies that θ^1 and θ^2 are identically zero, and thus rank$(S) = 0$, this is absurd. Consequently, there does not exist a system of 2-forms of degree two and rank 1.

The case rank$(S) = 2$

In these conditions we have $\dim(A(S)) = 1$. Denote by $\{e_1\}$ a basis of $A(S)$ that we extend to a basis $\{e_1, e_2, e_3\}$. In the dual basis $\{\omega^1, \omega^2, \omega^3\}$ of $\{e_1, e_2, e_3\}$ the forms θ^1 and θ^2 take the form

$$\begin{cases} \theta^1 = \sum C^1_{ij}\omega^i \wedge \omega^j, \\ \theta^2 = \sum C^2_{ij}\omega^i \wedge \omega^j. \end{cases}$$

We have $i(e_1)\theta^l = \sum_j C^l_{1j}\omega^j = 0$, then $C^l_{1k} = 0$ for all $l, k = 1, 2$. Thus

$$\begin{cases} \theta^1 = C^1_{23}\omega^2 \wedge \omega^3 \\ \theta^2 = C^2_{23}\omega^2 \wedge \omega^3 \end{cases}$$

the forms θ^1 and θ^2 are linearly dependent in $\Lambda^2(\mathbb{R}^3)^*$, which contradicts the hypothesis. Then there does not exist a system of two forms of degree two of rank 2.

The case rank$(S) = 3$

For this case the system (S) is non-degenerate (it is of maximum rank). Suppose that $\theta^1 \neq 0$. We can find a basis $\{\omega^1, \omega^2, \omega^3\}$ of $(\mathbb{R}^3)^*$ such that $\theta^1 = \omega^1 \wedge \omega^3$. Set $\theta^2 = a\omega^1 \wedge \omega^3 + b\omega^2 \wedge \omega^3 + c\omega^1 \wedge \omega^2$. Since the system $\{\theta^1, \theta^2 - a\theta^1\}$ is algebraically equivalent to (S), we can assume that $a = 0$. The hypothesis rank$(S) = 3$ implies that $b \neq 0$ or $c \neq 0$. Suppose that $b \neq 0$. Then (S) is equivalent to $\{\theta^1, \theta^2/b\}$ which allows us to assume that $b = 1$. Thus $\theta^2 = \omega^2 \wedge \omega^3 + c\omega^1 \wedge \omega^2$ and the change of basis $\alpha^i = \omega^i$, $i = 1, 2$ and $\alpha^3 = \omega^3 + c\omega^1$ gives the model according to

$$\begin{cases} \theta^1 = \alpha^1 \wedge \alpha^3, \\ \theta^2 = \alpha^2 \wedge \alpha^3. \end{cases}$$

Such a system will be called a 2-symplectic exterior system, or, more generally a 2-symplectic system. We deduce the following proposition:

Proposition 3.1 *Every 2-system (S) of rank 3 in $\mathbb{R}^3$, is a 2-symplectic exterior system.*

In this case the system (S) possesses a maximal solution F, of dimension 2 defined by $F = \ker \alpha^3$.

3.1.3   Classification for $k = 3$, $n = 3$

Let $(S) = \{\theta^1, \theta^2, \theta^3\}$ be a 3-system, that is, the three forms θ^i are independent in $\Lambda^2(\mathbb{R}^3)^*$. If the rank of (S) is 3, each of the 2—systems $\{\theta^1, \theta^2\}, \{\theta^1, \theta^3\}, \{\theta^2, \theta^3\}$ is 2-symplectic. Indeed, according to the above considerations, if this were not the case the forms $\theta^1, \theta^2, \theta^3$ would be independent. The system $\{\theta^1, \theta^2\}$ is 2-symplectic; then there exists a basis $\{\omega^1, \omega^2, \omega^3\}$ of $(\mathbb{R}^3)^*$ such that:

$$\begin{cases} \theta^1 = \omega^1 \wedge \omega^3, \\ \theta^2 = \omega^2 \wedge \omega^3, \end{cases}$$

and

$$\theta^3 = a\omega^1 \wedge \omega^3 + b\omega^2 \wedge \omega^3 + c\omega^1 \wedge \omega^2.$$

The system $\{\theta^1, \theta^2, \theta^3 - a\theta^1 - b\theta^2\}$ is algebraically equivalent to (S). Then we can suppose that $\theta^3 = c\omega^1 \wedge \omega^2$. By hypothesis $c \neq 0$, then we can take $c = 1$ to obtain the following model

$$\begin{cases} \theta^1 = \omega^1 \wedge \omega^3, \\ \theta^2 = \omega^2 \wedge \omega^3, \\ \theta^3 = \omega^1 \wedge \omega^2. \end{cases}$$

3.1.4   Classification for $k = 3$, $n = 4$, rank$(S) = 4$.

Maximal solutions

Let $(S) = \{\theta^1, \theta^2, \theta^3\}$ be an exterior system of rank 4 in $\mathbb{R}^4$, the exterior forms θ^i are of degree 2.

Proposition 3.2 *The dimension of all maximal solution is less than 3.*

As a matter of fact, for the opposite case the rank of (S) would be zero.

Systems equipped with a maximal solution of dimension 3

Let H be a maximal solution of (S) of basis $\{e_1, e_2, e_3\}$ which we extend to a basis $\{e_1, e_2, e_3, e_4\}$ of $\mathbb{R}^4$ with the dual basis $\{\omega^1, \omega^2, \omega^3, \omega^4\}$. But H is a solution of this system, therefore we have:

$$\theta^i = \sum C^i_{j4} \omega^j \wedge \omega^4.$$

Since the rank of the system is maximum, the rank of the matrix $\left(C^i_{j4} \right)$ is equal to 3. We deduce that the system is equivalent to

$$
\begin{cases}
\theta^1 = \omega^1 \wedge \omega^4, \\
\theta^2 = \omega^2 \wedge \omega^4, \\
\theta^3 = \omega^3 \wedge \omega^4.
\end{cases}
$$

Such a system is 3-symplectic in $\mathbb{R}^4$.

Case where the solution is 1-dimensional

Suppose that each form $\theta = \sum a_i \theta^i$ of this system (S) is of maximum rank. In this case all solutions of this system are of dimension 1. Otherwise the existence of a maximal solution of dimension 2 proves that the associated ideal to the system is 2-dimensional. Then there exist two linear forms $\{\omega^1, \omega^2\}$, such that

$$
\theta^i = \omega^1 \wedge \beta^i_1 + \omega^2 \wedge \beta^i_2.
$$

The rank of the system $(\beta^i_1, \beta^i_2)_{i=1,2,3}$ is less than 2, hence there exists i such that $\beta^i_1 = a\beta^j_1 + b\beta^k_1$, where $i \neq j$ and $i \neq k$. The rank of the form $\theta = \theta^i - a\theta^j - b\theta^k$ is less than 2, which contradicts the hypothesis. Hence all solutions are 1-dimensional. The problem of classification of such systems consists in classifying a system of skew-symmetric forms of rank 4, such that every combination is also of rank 4 (this problem is related to the determination of the number of independent vector fields in all points on the 3-dimensional sphere). Then,

$$
\begin{cases}
\theta^1 = \omega^1 \wedge \omega^4 + \omega^2 \wedge \omega^3, \\
\theta^2 = \omega^2 \wedge \omega^4 + \omega^3 \wedge \omega^1, \\
\theta^3 = \omega^3 \wedge \omega^4 + \omega^1 \wedge \omega^2.
\end{cases}
$$

Remark 3 *The 3-symplectic system which we have determined appears as the 'simple' system of maximum rank. Any other system of this nature can be seen as a deformation of the latter.*

3.2   k-symplectic exterior systems

Let E be an $n(k+1)$-dimensional vector space over a commutative field $\mathbb{K}$ of characteristic $\neq 2$, let F be an n-codimensional subspace of E and let $\theta^1, \cdots, \theta^k$ be k exterior 2-forms on E.

Definition 3.2 *We say that* $\theta^1, \cdots, \theta^k$ *is a k-symplectic system associated to F (or that the (k+1)-tuple* $\left(\theta^1, \cdots, \theta^k; F\right)$ *defines a k-symplectic structure on E) if the following conditions are fulfilled :*

1. *the exterior system* $\{\theta^1, \cdots, \theta^k\}$ *is non-degenerate, that is,*

$$A(\theta^1) \cap \cdots \cap A(\theta^k) = \{0\};$$

2. $\theta^p(x, y) = 0$ *for all* $x, y \in F$ *and* $p = 1, ..., k.$

Recall that the associated space to the alternating form θ is the set

$$A(\theta) = \{x \in E \mid \theta(x, y) = 0 \text{ for all } y \in E\}.$$

A subspace E_1 of E is called totally isotropic with respect to the form θ if it is a solution of the equation $\theta = 0$, that is

$$\theta(x, y) = 0$$

for all x, $y \in E_1$. Thus, the subspace F is totally isotropic with respect to the form θ^p for all $p = 1, ..., k$.

3.2.1   Solutions of k-symplectic exterior system

Let (S) be a k-symplectic exterior system (the subspace F is fixed to avoid all reference to this subspace). A subspace H of E is a solution of (S) if

$$\theta^p(x, y) = 0$$

for all $x, y \in H$. By the definition of the k-system (S) the subspace F is a maximal solution of (S).

3.2.2   Examples

1. Considering the real vector space $\mathbb{R}^{n(k+1)}$ endowed with its canonical basis $(e_{pi}, e_i)_{1 \leq p \leq k, 1 \leq i \leq n}$. We denote by $(\omega^{pi}, \omega^i)_{1 \leq p \leq k, 1 \leq i \leq n}$ the dual basis and let F be the subspace of $\mathbb{R}^{n(k+1)}$ spanned by the vectors $(e_{pi})_{1 \leq p \leq k, 1 \leq i \leq n}$. The system

$$\begin{cases} \theta^1 = \sum_{j=1}^n \omega^{1j} \wedge \omega^j, \\ \quad \vdots \\ \theta^k = \sum_{j=1}^n \omega^{kj} \wedge \omega^j \end{cases}$$

defines a k-symplectic exterior system on $\mathbb{R}^{n(k+1)}$ associated with F. This last structure will be called *the canonical k- symplectic structure of* $\mathbb{R}^{n(k+1)}$.

2. A symplectic vector space is a vector space of even dimension equipped with an exterior 2-form σ of maximum rank. A Lagrangian subspace of E is a maximal solution of the exterior equation $\sigma = 0$. Such a solution exists, and it is n-dimensional.

Let $(E^1, \sigma^1), ..., (E^k, \sigma^k)$ be symplectic vector spaces and let L^p be a Lagrangian subspace of E^p. Every quotient space E^p/L^p is of dimension n, there exists a vector subspace B of dimension n and a surjective linear mapping $\pi^p : E^p \longrightarrow B$ such that

$$\ker \pi^p = L^p$$

for all p ($p = 1, ..., k$). Let $(E^1 \times ... \times E^k, \sigma)$ be the product symplectic space where σ is the symplectic form defined by:

$$\sigma((x_1, ..., x_k), (y_1, ..., y_k)) = \sigma^1(x_1, y_1) + ... + \sigma^k(x_k, y_k).$$

Let E be the vector subspace of $E^1 \times ... \times E^k$ defined by:

$$E = \left\{ (x_1, ..., x_k) \in E^1 \times ... \times E^k \mid \pi^1(x_1) = ... = \pi^k(x_k) \right\}.$$

The linear mapping $\pi : E \longrightarrow B$ such that

$$\pi(x_1, ..., x_k) = \pi^1(x_1) = ... = \pi^k(x_k)$$

for each $(x_1, ..., x_k) \in E$, is surjective and satisfies

$$\ker \pi = L^1 \times ... \times L^k.$$

Let $i : E \longrightarrow E^1 \times ... \times E^k$ be the canonical injection and pr^u ($u = 1, ..., k$), the canonical projection $E^1 \times ... \times E^k \longrightarrow E^u$. For every u the composite $pr^u \circ i$ is the restriction of pr^u to E; it is surjective linear mapping. It is clear that the space E is of dimension $n(k+1)$. For every u ($u = 1, ..., k$), we take $\theta^u = (pr^u \circ i)^* \sigma^u$. Thus $(\theta^1, ..., \theta^k; \ker \pi)$ is a k-symplectic exterior system on E.

3.2.3 Classification of k-symplectic exterior systems

Theorem 3.1 *Let* $(\theta^1, ..., \theta^k; F)$ *be a k-symplectic exterior system on E. Then there exists a basis* $(w^{pi}, w^i)_{1 \leq p \leq k, 1 \leq i \leq n}$ *of E^* such that:*

$$\theta^p = \sum_{j=1}^{n} w^{pj} \wedge w^j$$

and

$$F = \ker w^1 \cap \cdots \cap \ker w^k.$$

Proof.   Let $\{f_1, \ldots, f_{nk}, g_1, \ldots, g_n\}$ be a basis of E such that F is spanned by the system $\{f_1, \ldots, f_{nk}\}$, and let $\{\gamma^1, \ldots, \gamma^{nk}, \omega^1, \ldots, \omega^n\}$ be its dual basis. The second condition of the definition of a k-symplectic exterior system allows us to see that the bilinear forms θ^p take the following form:

$$\theta^p = \sum_{j=1}^{n} \left(\sum_{i=1}^{nk} b_i^{pj} \gamma^i + \sum_{i=1}^{n} c_i^{pj} \omega^i \right) \wedge \omega^j,$$

where $b_i^{pj}, c_i^{pj} \in \mathbb{K}$. For all p, j ($p = 1, \ldots, k$ and $j = 1, \ldots, n$) we take

$$\omega^{pj} = \sum_{i=1}^{nk} b_i^{pj} \gamma^i + \sum_{j=1}^{n} c_j^{pj} \omega^j \quad , \quad \mu^{pj} = \sum_{i=1}^{nk} b_i^{pj} \gamma^i.$$

The set $\left\{ \mu^{pj} \mid p = 1, \ldots, k \text{ and } j = 1, \ldots, n \right\}$ is a basis of the dual space F^* of F. In fact, if an element $x \in F$ satisfies $\mu^{pj}(x) = 0$, for all p and j. Then the linear forms $i(x)\theta^p$ vanish identically, and it followss from the non-degeneracy of $\{\theta^1, \ldots, \theta^k\}$ that $x = 0$. The linear forms μ^{pj} are independent in F^*, and consequently they form a basis of F^*. The linear forms ω^{pj} are independent in E^*, the system

$$(\omega^{pi}, \omega^i)_{1 \leq p \leq k, \, 1 \leq i \leq n}$$

is a basis of E^*, and we have:

$$\theta^p = \sum_{j=1}^{n} \omega^{pj} \wedge \omega^j.$$

Let $(e_{pi}, e_i)_{1 \leq p \leq k, 1 \leq i \leq n}$ be a basis of E having $(\omega^{pi}, \omega^i)_{1 \leq p \leq k, 1 \leq i \leq n}$ as its dual basis. The bases $(f_i, g_t)_{1 \leq i \leq nk, 1 \leq t \leq n}$ and $(e_{pi}, e_i)_{1 \leq p \leq k, 1 \leq i \leq n}$ are related by:

$$f_i = \sum_{p=1}^{k} \left(\sum_{j=1}^{n} b_i^{pj} e_{pj} \right) \quad , \quad g_{i'} = \sum_{p=1}^{k} \left(\sum_{j=1}^{n} c_{i'}^{pj} e_{pj} \right) + e_{i'}.$$

This proves, in particular, that the vectors e_{pj} belong to F, and that $F = \ker \omega^1 \cap \cdots \cap \ker \omega^k$. ∎

Corollary 3.1 *The subspace F is a solution of (S).*

Definition 3.3 *The basis $(e_{pi}, e_i)_{1 \leq p \leq k, 1 \leq i \leq n}$ of E having as its dual basis $(\omega^{pi}, \omega^i)_{1 \leq p \leq k, 1 \leq i \leq n}$, is called a k-symplectic basis of E.*

Let F^p be the subspace of E defined by:

$$F^p = \bigcap_{q \neq p} A(\theta^q).$$

Proposition 3.3 *In the previous hypothesis and notations, we have:*

1. $F = F^1 \oplus \ldots \oplus F^k$ *(direct sum),*

2. *For every* p *($p= 1, \ldots, k$), the mapping*

$$i_p : x \longmapsto i(x)\theta^p$$

   defines an isomorphism of vector spaces from F^p *onto the annihilator* $\mathrm{Ann}(F)$ *of* F.

It is an immediate consequence of the theorem of classification. Recall that the annihilator $\mathrm{Ann}(F)$ is given by:

$$\mathrm{Ann}(F) = \{f \in E^* \mid f(x) = 0 \quad \forall x \in F\}.$$

This vector space is isomorphic to $(E/F)^*$. With respect to the $k-$symplectic basis $(e_{pi}, e_i)_{1 \leq p \leq k, 1 \leq i \leq n}$ of E and the dual basis $(\omega^{pi}, \omega^i)_{1 \leq p \leq k, 1 \leq i \leq n}$, we have :

1. $\mathrm{Ann}(F)$ is spanned by the linear forms $\omega^1, \ldots, \omega^n$,

2. F^p is spanned by the vectors $e_{p1}, \ldots, e_{pn}$.

Definition 3.4 *The subspaces* $F^1, \ldots, F^k$ *are called the characteristic spaces of the* $k-$*symplectic exterior system.*

For each $p = 1, \ldots, k$ we take

$$G^p = F^p \oplus \frac{E}{F}.$$

We have

$$\theta^p(x, y + f) = \theta^p(x, y)$$

for every $x \in F^p$, $y \in E$ and $f \in F$, then, $\theta^p(x, y)$ depends only on the coset $\bar{y}$ of y modulo F; this allows us to set

$$\bar{\theta}^p(x + \bar{y}, x' + \bar{y'}) = \theta^p(x, y') - \theta^p(x', y).$$

$\bar{\theta}^p$ defines a symplectic structure on $G^p = F^p \oplus E/F$.

3.3 k-symplectic endomorphisms

Let E be a vector space of dimension $n(k+1)$ over $\mathbb{K}$ and let $(\theta^1, \ldots, \theta^k; F)$ be a k-symplectic exterior system on E.

Definition 3.5 *Let f be an endomorphism of E. We say that it preserves the $k-$symplectic exterior system $(\theta^1, \ldots, \theta^k; F)$ if it leaves invariant both the system of exterior forms $\{\theta^1, \ldots, \theta^k\}$ and the subspace F, in other words, if the following conditions are satisfied :*

1. *$f(F) \subseteq F$,*

2. *There is $\sigma \in \mathfrak{S}_k$, such that*

$$\theta^p(f(x), f(y)) = \theta^{\sigma(p)}(x, y)$$

for all $p\,(p= 1, \ldots, k)$ and $x, y \in E$, where $\mathfrak{S}_k$ is the group of all permutations of the set $\{1, \ldots, k\}$. We denote by G the group of all automorphisms of the space E preserving the $k-$symplectic structure $(\theta^1, \ldots, \theta^k; F)$.

3.3.1   The group $Sp(k, n; E)$

Let $Sp(k, n; E)$ be the subgroup G defined by:

$$Sp(k, n; E) = \{f \in G \mid \theta^p(f(x), f(y)) = \theta^p(x, y), \text{ for all } x, \, y \in E \text{ and } p\} ,$$

$Sp(k, n; E)$ is called the $k-$symplectic group of E, and each element of $Sp(k, n; E)$ is called a $k-$symplectic automorphism of E.

Every element of G is a composite of elements of $Sp(k, n; E)$, and the isomorphisms $\tilde{\sigma}$ of E defined by

$$\tilde{\sigma}(e_{pi}) = e_{\sigma(p)i}, \quad \tilde{\sigma}(e_i) = e_i$$

for all $\sigma \in \mathfrak{S}_k$, $p = 1, \ldots, k$ and $i = 1, \ldots, n$, thus we can to return the study of G to $Sp(k, n; E)$.

Let $Sp(k, n; \mathbb{K})$ be the group of matrices of $k-$symplectic automorphisms of E, with respect to the $k-$symplectic basis.

Proposition 3.4 *$Sp(k, n; \mathbb{K})$ is the group of all matrices of the following type:*

$$\begin{pmatrix} T & 0 & \cdots & 0 & S_1 \\ 0 & T & \cdots & 0 & S_2 \\ \vdots & \vdots & \ddots & \vdots & \vdots \\ 0 & 0 & \cdots & T & S_k \\ 0 & 0 & \cdots & 0 & {}^t(T^{-1}) \end{pmatrix} \tag{3.1}$$

where T, $S_1, \cdots, S_k$ are $n \times n$ matrices, with coefficients in $\mathbb{K}$, such that, T is invertible and $T\,^t S_p = S_p\,^t T$, for each p $(p = 1, \ldots, k)$.

Proof. Let $f \in Sp(k, n; E)$, $p(p = 1, \ldots, k)$ and $j, j' = 1, \cdots, n$. Write

$$f(e_{j'}) = \sum_{i=1}^{n} t'^{i}_{j'}\, e_i + \sum_{q=1}^{k}\left(\sum_{i=1}^{n} t'^{qi}_{j'} e_{qi}\right),$$

where $t'^{i}_{j'}$, $t'^{qi}_{j'} \in \mathbb{K}$. Each element of $Sp(k, n; E)$ leaves invariant the vector subspace F; then we can write

$$f(e_{pj}) = \sum_{q=1}^{k}\sum_{i=1}^{n} t^{qi}_{pj} e_{qi}.$$

The relationship $\theta^p(f(e_{pj}), f(e_{j'})) = \theta^p(e_{pj}, e_{j'}) = \delta_{jj'}$ implies that

$$\sum_{i=1}^{n} t^{pi}_{pj} t'^{i}_{j'} = \delta_{jj'}.$$

Let T' (resp. T^q_p) be the matrix with coefficients t'^{i}_{j} (resp. t^{qi}_{pj}) where p, $q = 1, \ldots, k$. The matrices T' and T^q_p $(p = 1, \ldots, k)$ which are invertible, are related by:

$$T^p_p = {}^t(T'^{-1}).$$

For $q \neq p$ we have $\theta^q(f(e_{pj}), f(e_{j'})) = \theta^q(e_{pj}, e_{j'}) = 0$, then

$$\sum_{i=1}^{n} t^{qi}_{pj} t'^{i}_{j'} = 0.$$

Hence $T^q_p\,{}^t T' = 0$. But the matrix T' is invertible, thus the matrices $T^q_p = 0$. According to $\theta^p(f(e_j), f(e_{j'})) = \theta^p(e_j, e_{j'})$ we have

$$\sum_{i=1}^{n} t'^{pi}_{j} t'^{i}_{j'} = \sum_{i=1}^{n} t'^{i}_{j} t'^{pi}_{j'}.$$

Consequently the matrices S_p with coefficients $t'^{pi}_{j}(i, j = 1, \ldots, n)$ satisfy

$${}^t S_p T' = {}^t T' S_p$$

for each p $(p = 1, \ldots, k)$. Since $T = T^p_p = {}^t(T'^{-1})$, the matrix of f with respect to the k-symplectic basis is of the form 3.1. ∎

3.3.2   The Lie algebra $\mathfrak{sp}(k, n; E)$

We assume in this subsection that $\mathbb{K} = \mathbb{R}$ or $\mathbb{C}$. For this case the group $Sp(k, n; E)$ is a Lie group. The Lie algebra of this group, which is denoted by $\mathfrak{sp}(k, n; E)$ and identified below with the tangent space of this group in the identity mapping of E, is formed of the endomorphisms u of E according to:

$$u(F) \subseteq F, \quad \theta^p(u(x), y) + \theta^p(x, u(y)) = 0,$$

for all $x, y \in E$ and $p = 1, \ldots, k$.

Let $\mathfrak{sp}(k, n; \mathbb{K})$ be the Lie algebra of $Sp(k, n; \mathbb{K})$, which is the Lie algebra of all matrices of endomorphisms, with respect to the k-symplectic basis of E, belonging to the Lie algebra $\mathfrak{sp}(k, n; E)$.

Proposition 3.5 *The Lie algebra $\mathfrak{sp}(k, n; \mathbb{K})$ is constituted of the matrices of type*:

$$\begin{pmatrix} A & 0 & \cdots & 0 & S_1 \\ 0 & A & \cdots & 0 & S_2 \\ \vdots & \vdots & \ddots & \vdots & \vdots \\ 0 & 0 & \cdots & A & S_k \\ 0 & 0 & \cdots & 0 & -{}^tA \end{pmatrix}$$

where $A, S_1, \cdots, S_k$ are $n \times n$ matrices with coefficients in $\mathbb{K}$, with ${}^tS_p = S_p$, for every p ($p = 1, \ldots, k$).

Proof. Let $u \in \mathfrak{sp}(k, n; E)$, $p = 1, \ldots, k$ and $j = 1, \ldots, n$. Write

$$u(e_{pj}) = \sum_{q=1}^k \left(\sum_{h=1}^n c_{pj}^{qh} e_{qh} \right) + \sum_{h=1}^n c_{pj}^h e_h,$$

$$u(e_j) = \sum_{q=1}^k \left(\sum_{h=1}^n c_j^{qh} e_{qh} \right) + \sum_{h=1}^n c_j^h e_h.$$

The relationship $\theta^p(u(x), y) + \theta^p(x, u(y)) = 0$, for all $x, y \in E$, implies that:

$$c_{pi}^i = 0,$$
$$c_{qj}^{pi} = 0 \text{ if } p \neq q,$$
$$c_{pj}^{pi} + c_i^j = 0,$$
$$c_j^{pi} = c_i^{pj}.$$

The matrix of u with respect to the k-symplectic basis is of the expected form. ∎

Corollary 3.2 *The group $Sp(k, n; E)$ is of dimension $n^2 + \frac{kn(n+1)}{2}$.*

Corollary 3.3 *The k-symplectic group and its Lie algebra leave invariant the characteristic subspaces of the k-symplectic exterior system.*

In the hypothesis and notations of previous subsection we consider the space $G^p = F^p \oplus E/F$ equipped with its a symplectic structure defined by

$$\overline{\theta}^p(x + \overline{y}, x' + \overline{y'}) = \theta^p(x, y') - \theta^p(x', y)$$

for all $x + \overline{y}, x' + \overline{y'} \in G^p$, where $p = 1, \ldots, k$.

Corollary 3.4 *For every $f \in Sp(k, n; E)$ the association*

$$\overline{f}_p : x + \overline{y} \longmapsto f(x) + \overline{f(y)}$$

defines an element of the symplectic group $Sp(\overline{\theta}^{\,p}, G^{\,p})$ of the symplectic vector space $(G^p, \overline{\theta}^p)$.

Corollary 3.5 *If $(\theta^1, \ldots, \theta^k; F)$ is a k-symplectic exterior system on a vector space of dimension $n(k+1)$, then the forms $\theta^1, \ldots, \theta^k$ are of rank $2n$.*

3.4 k-symplectic geometry

3.4.1 k-symplectic orthogonality

Let E be an $n(k+1)$-dimensional vector space over a commutative field $\mathbb{K}$ with characteristic not equal to 2, equipped with a k-symplectic structure $(\theta^1, \ldots, \theta^k; F)$.

Definition 3.6   1. *Let x and y be two vectors of E. We say that x is orthogonal to y, with respect to the system $\{\theta^1, \cdots, \theta^k\}$, or, symbolically, $x \perp y$, if $\theta^p(x, y) = 0$ for all p ($p = 1, \ldots, k$).*

2. *Let L and M be two vector subspaces of E. We say that L is orthogonal to M with respect to the system $\{\theta^1, \cdots, \theta^k\}$, or, symbolically, $L \perp M$, if $x \perp y$ for all $x \in L$ and $y \in M$.*

The k-symplectic orthogonal of a non-empty set A of E is the vector subspace of E defined by:

$$A^\perp = \{x \in E \mid \theta^p(x, y) = 0, \ \forall y \in A, \ \forall p \ (p = 1, \ldots, k)\}.$$

Definition 3.7 *We say that E is an orthogonal sum of the subspace L and M of E if we have:*

$$E = L \oplus M , \ L = M^{\perp}.$$

Notice that we have:

1. The non-degeneracy of the system $\{\theta^1, \cdots, \theta^k\}$ is equivalent to $E^{\perp} = \{0\}$.

2. The 2-symplectic orthogonal $X^{\perp}$ of an element X of E is not necessarily an hyperplane. Considering, for instance, the real space $\mathbb{R}^3$ equipped with its canonical 2-symplectic structure defined by:

$$\begin{cases} \theta^1 = \omega^1 \wedge \omega^3, \\ \theta^2 = \omega^2 \wedge \omega^3, \end{cases}$$

   and $F = \ker \omega^3$, where $\{\omega^1, \omega^2, \omega^3\}$ is the dual basis of the canonical basis $\{e_1, e_2, e_3\}$ of E. The 2-symplectic orthogonal $e_1^{\perp}$ of e_1 is the hyperplane spanned by the system $\{e_1 , e_2\}$, while the 2-symplectic orthogonal $e_3^{\perp}$ of e_3 is the line $\mathbb{R}e_3$.

Proposition 3.6 *For all non-empty subsets A and B of E, we have:*

1. $A \subseteq B \implies B^{\perp} \subseteq A^{\perp}$,

2. $A \subseteq A^{\perp\perp}$.

Definition 3.8 *A subspace L of E will be called totally isotropic if $L \subseteq L^{\perp}$.*

A subspace L of E is a totally isotropic subspace if and only if, the following property is satisfied:

$$\theta^p(x, y) = 0$$

for all $x, y \in L$ and $p = 1, \ldots, k$. Every totally isotropic subspace is contained in a maximal totally isotropic subspace.

Definition 3.9 *A maximal totally isotropic subspace will be called a Lagrangian subspace of E.*

Proposition 3.7 *Every element of the* k*-symplectic group* $Sp(k, n; E)$ *transforms the totally isotropic subspaces (resp. Lagrangian subspaces) of* E *in totally isotropic subspaces (resp. Lagrangian subspaces) .*

Proof. Let $f \in Sp(k, n; E)$ and let L be a totally isotropic subspace of E. For any $x, y \in E$ we have:

$$\theta^p(f(x), f(y)) = \theta^p(x, y) = 0$$

for each $p = 1, \ldots, k$. The subspace $f(L)$ is totally isotropic.

Let L be a Lagrangian subspace of E. Suppose $f(L)$ is strictly contained in $f(L)^\perp$, then there exists $a' = f(a)$ belonging to $f(L)^\perp - f(L)$ such that for each $x \in f(L)$ we have $\theta^p(a, x) = 0$. Set $A = L \oplus \mathbb{K}a$. We have $\theta^p(u, v) = 0$ for all $u, v \in A$, and $A \subseteq A^\perp$. Consequently L is not maximal, which contradicts the hypothesis; then $f(L)$ is a Lagrangian subspace. ∎

Proposition 3.8 *Let* L *be a Lagrangian subspace of* E, *and* M_0 *a totally isotropic subspace of* E *supplementary to* L. *Then there exists a Lagrangian subspace* M *of* E *supplementary to* L *and containing* M_0.

Proof. Let M be a maximal element of the set of totally isotropic subspaces of E transverse to L and containing M_0. Then M is a Lagrangian subspace of E. ∎

Proposition 3.9 *For every vector subspace* L *of* E *the following properties are equivalent :*

1. L *is a Lagrangian subspace of* E,

2. $L = L^\perp$.

Proof. If L is strictly contained in $L^\perp$ there exists an element a of $E - L$ such that

$$\theta^p(a, y) = 0$$

for all $y \in L$ and $p = 1, \ldots, k$. Let $L' = L \oplus \mathbb{K}a$. Obviously, we have $L \subset L'$. Let $z = x + \lambda a \in L'$, with $x \in L$ and $\lambda \in \mathbb{K}$. We have

$$\begin{aligned}
\theta^p(z, z') &= \theta^p(x + \lambda a, x' + \lambda' a) \\
&= \theta^p(x, x') + \lambda \theta^p(x, a) + \lambda' \theta^p(a, x') \\
&= 0,
\end{aligned}$$

for all $z' = x' + \lambda' a$ belonging to L' ($x' \in L$ and $\lambda' \in \mathbb{K}$) and $p = 1, \ldots, k$, then $z \in L'^\perp$. Thus $L' \subset L'^\perp$, which contradicts the property that L is a maximal totally isotropic subspace of E. This proves that $1 \Longrightarrow 2$. The implication $2 \Longrightarrow 1$ is immediate. ∎

Proposition 3.10 *Let L be a vector subspace of E.*

1. *If L is a totally isotropic subspace, then* $\dim L \leq nk$.

2. *If L is Lagrangian then,* $n \leq \dim L \leq nk$.

Proof. Let L be a totally isotropic subspace of E, $\{f_1, \ldots, f_h\}$ be a basis of L which we extend to a basis of E denoted by $\{f_1, \ldots, f_h, u_1, \ldots, u_{m-h}\}$, and let $\{f_1^*, \ldots, f_h^*, u_1^*, \ldots, u_{m-h}^*\}$ be its dual basis. The two forms θ^p can be written as:

$$\theta^p = \sum_{j=1}^{m-h} \left(\sum_{i=1}^{h} a^{pi\ j} f_i^* + \sum_{i=1}^{m-h} b^{pij} u_i^* \right) \wedge u_j^*,$$

where $a^{pi\ j}, b^{pij} \in \mathbb{K}$. Consider the following matrix:

$$A = \begin{pmatrix} a^{11\ 1} & a^{12\ 1} & \cdots & a^{1h\ 1} \\ \vdots & \vdots & \cdots & \vdots \\ a^{11\ m-h} & a^{12\ m-h} & \cdots & a^{1h\ m-h} \\ \vdots & \vdots & \cdots & \vdots \\ a^{k1\ 1} & a^{k2\ 1} & \cdots & a^{kh\ 1} \\ \vdots & \vdots & \cdots & \vdots \\ a^{k1\ m-h} & a^{k2\ m-h} & \cdots & a^{kh\ m-h} \end{pmatrix}.$$

and show that it is of rank h. For all $X = (x_1, \cdots, x_h) \in \mathbb{K}^h$ the relationship $A^t X = 0$ implies that the vector $x_1 f_1 + \cdots + x_h f_h$ belongs to $A(\theta^1) \cap \cdots \cap A(\theta^k)$; the non-degeneracy of the system $\{\theta^1, \ldots, \theta^k\}$ implies that $X = 0$, thus the matrix A is of rank h. It follows that we have, in particular, $h \leq k(m - h)$, consequently $h \leq nk$ this proves the first assertion.

Suppose now that L is a Lagrangian subspace of E, and consider the matrix

$$B = \begin{pmatrix} a^{11\ 1} & a^{11\ 2} & \cdots & a^{11\ m-h} \\ \vdots & \vdots & \cdots & \vdots \\ a^{1h\ 1} & a^{1h\ 2} & \cdots & a^{1h\ m-h} \\ \vdots & \vdots & \cdots & \vdots \\ a^{k1\ 1} & a^{k1\ 2} & \cdots & a^{k1\ m-h} \\ \vdots & \vdots & \cdots & \vdots \\ a^{kh\ 1} & a^{kh\ 2} & \cdots & a^{kh\ m-h} \end{pmatrix}.$$

We show that B is of rank $m - h$. For each $Y = (y_1, \cdots, y_{m-h}) \in \mathbb{K}^{m-h}$, the relationship $B^tY = 0$ implies that the vector $y = \sum_{j=1}^{m-h} y_j u_j$ satisfies

$$\theta^p(y, f_i) = -\sum_{j=1}^{m-h} a^{pi\,j} y_j = 0$$

for any $p\,(p = 1, \ldots, k)$ and $i\,(i = 1, \ldots, h)$; then

$$\theta^p(y, g) = 0$$

for all $p(p = 1, \ldots, k)$ and $g \in L$. This proves that the vector y is in the k−symplectic orthogonal $L^\perp$ of L; but L is a Lagrangian subspace E and the vector y belonging to L, thus $Y = 0$. We have proved the following implication

$$B^tY = 0 \Longrightarrow Y = 0,$$

consequently the matrix B is of rank $m - h$; then $m - h \leq kh$ and $n \leq h$ which proves the second assertion. ■

Corollary 3.6 *The vector subspaces F and $G = Vect(e_1, \ldots, e_n)$ are maximal solutions of the system*

$$\begin{cases} \theta^1 = 0, \\ \quad \vdots \\ \theta^k = 0. \end{cases}$$

We observe that we have :

1. For $k = 1$ we find the classic case in which all Lagrangian subspaces are n-dimensional.

2. The vector subspaces F and $G = \mathrm{Vect}(e_1, \ldots, e_n)$ are Lagrangian subspaces of E of dimensions nk and n respectively. There also exists Lagrangian subspaces whose the dimensions are strictly contained between n and nk. Consider, for example, in the case where $n = k = 2$, the vector subspace L of E spanned by $e_{11} + e_1$, e_{12} and e_{22}; L is a Lagrangian subspace of E of dimension 3.

3.4.2   Adjoint endomorphisms

Let E be a vector space of dimension $n(k+1)$ and let $(\theta^1, \ldots, \theta^k; F)$ be a $k-$symplectic exterior system on E. Let $\tilde{\eta}$ be the mapping of E into the $\mathbb{K}-$vector space $\hom(E, \mathbb{K}^k)$ defined by:

$$\tilde{\eta}(x)(y) = (\theta^1(x, y), \ldots, \theta^k(x, y))$$

for all $x, y \in E$. The non-degeneracy of the system $\{\theta^1, \ldots, \theta^k\}$ implies that the mapping $\tilde{\eta}$ is injective.

For each endomorphism u of E, we can associate a linear mapping from $\hom(E, \mathbb{K}^k)$ into itself, denoted by ^tu and called transpose of u, defined by:

$$^tu(\xi) = \xi \circ u$$

for every $\xi \in \hom(E, \mathbb{K}^k)$.

Proposition 3.11 *For all $u, v \in End(E)$ and $\lambda \in \mathbb{K}$, we have :*

1. $^t(u + v) = {}^tu + {}^tv, \qquad {}^t(\lambda u) = \lambda {}^tu;$

2. $^t(u \circ v) = {}^tv \circ {}^tu;$

3. ^tid_E *is the identity mapping of* $\hom(E, \mathbb{K}^k);$

4. *if $u \in GL(E)$ then $^tu \in GL(\hom(E, \mathbb{K}^k))$ and $(^tu)^{-1} = {}^t(u^{-1})$.*

Let u be an endomorphism of E, satisfying the hypothesis:

$$\text{Im } {}^tu \circ \tilde{\eta} \subseteq \text{Im } \tilde{\eta}. \tag{3.2}$$

For each $t \in E$ there exists an unique $t' \in E$, such that

$$({}^tu \circ \tilde{\eta})(t) = \tilde{\eta}(t').$$

It is clear that the association $t \longmapsto t'$ defines a mapping from E into itself, denoted by u^*, such that for all t and x belonging to E we have:

$$(\theta^1(t, u(x)), \ldots, \theta^k(t, u(x))) = (\theta^1(u^*(t), x), \ldots, \theta^k(u^*(t), x)).$$

Consequently, we deduce that for all $u, v \in End(E)$ satisfying the relationship 3.2 and $\lambda \in \mathbb{K}$, we have the following properties :

1. $\theta^p(t, u(x)) = \theta^p(u^*(t), x)$ for all $p(p = 1, \ldots, k)$ and $t, x \in E$;

2. the mapping u^* is linear;

3. $(u+v)^* = u^* + v^*$, $(\lambda u)^* = \lambda u^*$;

4. $(u \circ v)^* = v^* \circ u^*$, $u^{**} = u$, $(id_E)^* = id_E$;

5. if $u \in GL(E)$ then $u^* \in GL(E)$ and we have $(u^*)^{-1} = (u^{-1})^*$.

Conversely, let u be an element of $\text{End}(E)$ in such way that there exists an endomorphism u^* of E satisfying:

$$\theta^p(t, u(x)) = \theta^p(u^*(t), x)$$

for all p $(p = 1, \ldots, k)$ and $t, x \in E$; then we have ${}^t u(\tilde{\eta}(t))$. Furthermore:

Proposition 3.12 *Let u be an endomorphism of E. The following properties are equivalent :*

1. *There exists an endomorphism u^* of E such that:*

$$\theta^p(t, u(x)) = \theta^p(u^*(t), x),$$

   for all $p(p = 1, \ldots, k)$ and $t, x \in E$,

2. $\text{Im } {}^t u \circ \tilde{\eta} \subseteq \text{Im } \tilde{\eta}$.

Definition 3.10 *If an endomorphism u of E satisfies one of the equivalent conditions of the previous proposition, the endomorphism u^* defined above is called an adjoint endomorphism of u.*

Remark 4 *For $k = 1$ the mapping $\tilde{\eta}$ is an isomorphism of vector spaces from E into E^*, and the relation*

$$\text{Im } {}^t u \circ \tilde{\eta} \subseteq \text{Im} \tilde{\eta} = E^*$$

is satisfied for any endomorphism u of E; hence, for each endomorphism of E, there is, in this case associated, an adjoint endomorphism. This situation is not automatic when $k \geq 2$, because $\text{Im } \tilde{\eta}$ is of dimension $n(k+1)$; it is then strictly contained in $\text{hom}(E, \mathbb{K}^k)$, which is $kn(k+1)$-dimensional.

Proposition 3.13 *We assume $k \geq 2$. For every endomorphism u of E the following properties are equivalent:*

1. u *admits an adjoint endomorphism;*

2. $u(F^p) \subseteq F^p$ *for any* $p(p = 1, \ldots, k)$ *and* $u_{|F^1} = \cdots = u_{|F^k};$

3. *The matrix of* u *with respect to the* $k-$*symplectic basis of* E *has the form:*

$$
\begin{pmatrix}
A & & 0 & Q_1 \\
& \ddots & & \vdots \\
0 & & A & Q_k \\
0 & & 0 & B
\end{pmatrix}
\tag{3.3}
$$

where $A, B, Q_1, \ldots, Q_k$ *are* $n \times n$ *matrices of which coefficients are in* $\mathbb{K}$.

Proof. We first express the matrix of u with respect to the $k-$symplectic basis $(e_{pi}, e_i)_{1 \leq p \leq k, 1 \leq i \leq n}$ of E, knowing that u satisfies the condition

$$
\mathrm{Im}(^t u \circ \tilde{\eta}) \subseteq \mathrm{Im}\ \tilde{\eta}.
$$

Denote by $(f_w)_{1 \leq w \leq k}$ the canonical basis of $\mathbb{K}^k$. The elements E_w^b of $\mathrm{hom}(E, \mathbb{K}^k)$ given by:

$$
E_w^b(e_a) = \delta_a^b f_w,
$$

where a and $b \in \{11, \ldots, pj, \ldots, kn, 1, \ldots, j, \ldots n \}$ and $w \in \{1, \ldots, k\}$, form a basis of $\mathrm{hom}(E, \mathbb{K}^k)$. Set

$$
\tilde{\eta}(e_a) = \sum_{b,\, w} \gamma_{ab}^w E_w^b.
$$

If $a = pi \ (p = 1, \ldots, k; i = 1, \ldots, n)$ we have

$$
0 = \tilde{\eta}(e_{pi})(e_{qj}) = \sum_{b,\, w} \gamma_{pib}^w E_w^b(e_{qj}) = \sum_{b,\, w} \gamma_{pjb}^w \delta_{qj}^b f_w = \sum_{w} \gamma_{pi\ qj}^w f_w
$$

for all $q, j\ (q = 1, \ldots, k; j = 1, \ldots, n)$, then

$$
\gamma_{pi\ qj}^w = 0
$$

for all w, p, q, i, j. Similarly, we have:

$$
\delta_{ij} f_p = \tilde{\eta}(e_{pi})(e_j) = \sum_{b,\, w} \gamma_{pib}^w E_w^b(e_j) = \sum_{b,\, w} \gamma_{pib}^w \delta_j^b f_w = \sum_{w} \gamma_{pi\ j}^w f_w,
$$

thus,

$$
\gamma_{pi\ j}^w = \delta_p^w \delta_{ij}
$$

for all p, w, i, j; thus

$$\tilde{\eta}(e_{pi}) = \sum_w \gamma^w_{pii} E^i_w = E^i_p.$$

If $a = i$ $(i = 1, \ldots, n)$ we have

$$\tilde{\eta}(e_i) = \sum_{b,\,w} \gamma^w_{ib} E^b_w$$

where $b = \ldots pj \ldots, \ldots j \ldots$ $(p = 1, \ldots, k; j = 1, \ldots, n)$; hence

$$0 = \tilde{\eta}(e_i)(e_j) = \sum_{b,\,w} \gamma^w_{ib} E^b_w(e_j) = \sum_{b,\,w} \gamma^w_{ib} \delta^b_j f_w = \sum_w \gamma^w_{i\,j} f_w,$$

thus,

$$\gamma^w_{ij} = 0$$

for all w, i, j. From the equalities

$$\tilde{\eta}(e_i)(e_{qj}) = \sum_{b,\,w} \gamma^w_{ib} E^b_w(e_{qj}) = \sum_{b,w} \gamma^w_{ib} \delta^b_{qj} f_w = \sum_w \gamma^w_{i\,qj} f_w$$

$$= -\tilde{\eta}(e_{qj})(e_i) = -\delta_{ij} f_q$$

it follows that

$$\gamma^w_{i\,qj} = -\delta_{ij}\delta^w_q = -\gamma^w_{qj\,i}.$$

Consequently,

$$\tilde{\eta}(e_i) = \sum_{q,\,w,j} \gamma^w_{i\,qj} E^{qj}_w = -\sum_{q,\,w,j} \delta_{ij}\delta^w_q E^{qj}_w = -\sum_{q,\,w} \delta^w_q E^{qi}_w = -\sum_{p=1}^{k} E^{pi}_w = \Omega^i.$$

This proves that Im $\tilde{\eta}$ is spanned by the linear mapping:

$$\Omega^i_p = E^i_p,$$

$$\Omega^i = -\sum_{p=1}^{k} E^{pi}_p,$$

where $p = 1, \ldots, k$, $i = 1, \ldots, n$. We have proved that for an endomorphism u of E the following properties are equivalent:

1. There exists an endomorphism u^* of E satisfying

$$\theta^p(t, u(x)) = \theta^p(u^*(t), x)$$

for all $p(p = 1, \ldots, k)$ and $t, x \in E$;

2. Im $({}^t u \circ \tilde\eta)$ is contained in the vector subspace of $\hom(E, \mathbb{K}^k)$ spanned by Ω_p^i and Ω^i ($p = 1, \ldots, k$, $i = 1, \ldots, n$).

Write

$$u(e_{pi}) = \sum_{q=1}^{k}(\sum_{j=1}^{n} t_{pi}^{qj} e_{qj}) + \sum_{j=1}^{n} t_{pi}^{j} e_j$$

and

$$u(e_i) = \sum_{q=1}^{k}(\sum_{j=1}^{n} t_i^{qj} e_{qj}) + \sum_{j=1}^{n} t_i^{j} e_j.$$

We have the following relationships:

$$\begin{cases} ({}^t u \circ \tilde\eta)(e_{qi}) = \sum_{p=1}^{k}\left(\sum_{j=1}^{n} t_{pj}^{i} E_q^{\,pj}\right) + \sum_{j=1}^{n} t_j^{i} E_q^{j} \\ ({}^t u \circ \tilde\eta)(e_i) = -\sum_{p,\,q,j} t_{pj}^{qi} E_q^{\,pj} - \sum_{j,\rho} t_j^{\rho s} E_\rho^{j}. \end{cases}$$

The endomorphism $({}^t u \circ \tilde\eta)(e_{qi})$ belongs to Im $\tilde\eta$ if and only if

$$\sum_{p=1}^{k}\left(\sum_{j=1}^{n} t_{pj}^{i} E_q^{\,pj}\right)$$

is a linear combination of Ω^i; the expression $\sum_{p=1}^{k}(\sum_{j=1}^{n} t_{pj}^{i} E_q^{\,pj})$ is thus reduced to the single term:

$$\sum_{p=1}^{k}\left(\sum_{j=1}^{n} t_{pj}^{i} E_q^{\,pj}\right) = t_{qj}^{i} E_q^{\,qj}.$$

The hypothesis $k \geq 2$ implies that $t_{qj}^{i} = 0$.

For all i ($i = 1, \ldots, n$) ,the endomorphism $({}^t u \circ \tilde\eta)(e_i)$ belongs to Im $\tilde\eta$ if and only if $\sum_{p,\,q,j} t_{pj}^{qi} E_q^{\,pj}$ is a linear combination of Ω^i . The expression $\sum_{p,\,q,j} t_{pj}^{qi} E_q^{\,pj}$ can be written in the form:

$$\sum_{p,\,q,j} t_{pj}^{qi} E_q^{\,pj} = \sum_{p,j} t_{pj}^{pi} E_p^{\,pj}.$$

In these conditions we have

$$\sum_{p,j} t_{pj}^{pi} E_{\,p}^{\,pj} = \sum_{i'=1}^{n} \lambda_{i'} \Omega^{i'},$$

thus,

$$\sum_{p,i} t^{pi}_{pi} E^{pi}_p(e_{qj}) = \sum_{p,i} t^{pi}_{pi} \delta^{pi}_{qj} f_p = t^{qi}_{qj} f_q$$

and

$$\sum_{i=1}^{n} \lambda_{i'} \Omega^{i'}(e_{qj}) = -\sum_{i=1}^{n} \lambda_{i'} \sum_{p=1}^{k} E^{pi'}_p(e_{qj}) = -\sum_{i=1}^{n} \lambda_{i'} \sum_{p=1}^{k} \delta^{pi'}_{qj} f_p = -\lambda_j f_q.$$

Thus $t^{pi}_{pi'} = a^i_{i'} = -\lambda_{i'}$ does not depend on p ($p = 1, \ldots, k$). Consequently we have

$$u(e_{pi}) = \sum_{j=1}^{n} a^j_i e_{pj}$$

and

$$u(e_i) = \sum_{p=1}^{k} \left(\sum_{j=1}^{n} t^{pj}_i e_{pj} \right) + \sum_{j=1}^{n} t^j_i e_j.$$

We have shown that the matrix of u with respect to the k-symplectic basis can be written in the expected form. ∎

Corollary 3.7 *If u belongs to the k−symplectic group $Sp(k, n; E)$ then u^* exists and we have $u^* = u^{-1}$.*

3.5 k-symplectic transvections

The study of symmetries of a given structure (it is the mappings that preserve this structure) permits a fine study of the underlying geometry. There are detailed studies on generators of the linear and symplectic groups in the monographs by Artin [1] and Dieudonné [17], making obvious the role played by dilations and transvections. These transformations generate the linear group if the field $\mathbb{K}$ is different from $\mathbb{F}_2$ (the field having two elements). If E is a vector space over $\mathbb{F}_2$, then the linear group $GL_{\mathbb{K}}(E)$ coincides with the special linear group $SL_{\mathbb{K}}(E)$. As for the symplectic group $Sp(\theta, E)$ relative to a given symplectic structure θ, it is generated by the set of its symplectic transvections.

In accordance with the classic study corresponding to the symplectic case, we study in this section the k−symplectic transvections, that is transvections preserving a k−symplectic exterior system $(\theta^1, \ldots, \theta^k; F)$ on a $\mathbb{K}$-vector space E, in order to see their role in the study of generators of $Sp(k, n; E)$.

In contrast to the classic case of a symplectic structure, the k−symplectic transvections do not generate the group $Sp(k, n; E)$; they generate a normal subgroup of $Sp(k, n; E)$ denoted $Tp(k, n; E)$. For $k = 1$ the 1-symplectic group $Sp(1, n; E)$ does not coincide with $Sp(n, E)$; in fact, $Sp(1, n, E)$ is the subgroup of $Sp(n, E)$ formed by elements which leave invariant the symplectic structure θ (of E) and which also leave a Lagrangian subspace L invariant. The transvections of $Sp(1, n, E)$ are therefore the transvections of $Sp(n, E)$ which leave this Lagrangian subspace invariant. It is surprising to see that the subgroup $Tp(1, n, E)$ spanned by these transvections appears naturally within the framework of the geometrical quantification of Kostant - Souriau and equally, in Kirillov's approach to the harmonic analysis in nilpotent Lie groups.

3.5.1   Dilations and transvections

Let E be a vector space over a commutative field $\mathbb{K}$, let H be a vector hyperplane of E, and let u be an endomorphism of E fixing every element of H. If x is a vector of E not belonging to H we can write

$$u(x) = \gamma x + a,$$

where $a \in H$; the scalar γ is independent of the chosen vector $x \notin H$.

If $\gamma \neq 1$ then γ is a characteristic value of u and $E(\gamma; u)$ is a line S, which supplements $E(1; u) = H$, where $E(\gamma; u)$ denotes the set of elements $z \in E$ such that $u(z) = \gamma z$. The only vector lines D such that $u(D) \subset D$ are the lines contained in H and the line S. We say that u is a *dilation of ratio* γ, *with line* S, *and with fixed hyperplane* H. A subspace V satisfies $u(V) \subset V$ if and only if either $V \subset H$ or $S \subset V$.

Suppose $\gamma = 1$. Let $\varphi(x) = 0$ be an equation of H (φ belongs to the dual E^* of the space E), then there exists an unique vector $t \in H$ such that

$$u(x) = x + \varphi(x)t$$

for all $x \in E$. In this case we say that u (which is a bijection) is *a transvection with fixed hyperplane* H.

When $u \neq \mathrm{id}_E$ the line $T = \mathbb{K}t$ is independent of the particular equation taken for H, and we say that u is *a transvection with line* T *and hyperplane* H. The scalar 1 is the only characteristic value of u (if $H \neq \{0\}$) and $E(1, u) = H$, if $u \neq \mathrm{id}_E$.

If $H \neq \{0\}$ and $u \neq \mathrm{id}_E$ the only vector lines D such that $u(D) = D$ are the lines in H; a necessary and sufficient condition for a subspace V to be mapped into itself is then that either $V \subset H$ or $T \subset V$.

Proposition 3.14 *Let v be an automorphism of E. We have:*

1. *If u is a dilation with ratio γ with line S and fixed hyperplane H then vuv^{-1} is a dilation with ratio γ with line $v(S)$ and fixed hyperplane $v(H)$;*

2. *If u is a transvection with line T and hyperplane H, then vuv^{-1} is a transvection with line $v(T)$ and hyperplane $v(H)$.*

Let $\Gamma(E, H)$ be the set of automorphisms of E leaving the hyperplane H pointwise fixed:

$$\Gamma(E, H) = \{\sigma \in GL(E) \mid \sigma(h) = h, \ \forall h \in H\},$$

and let $\Theta(E, H)$ be a set of transvections with hyperplane H, then we have

Proposition 3.15 *$\Theta(E, H)$ is a normal abelian subgroup of $\Gamma(E, H)$ which isomorphic to additive group of H. In particular the inverse of the transvection*

$$x \longmapsto x + \varphi(x)t,$$

where $\varphi(x) = 0$ is an equation of the hyperplane and $t \in H$, is the transvection

$$x \longmapsto x - \varphi(x)t.$$

In fact, consider the mapping f from $\Gamma(E, H)$ onto $\mathbb{K}^*$ defined by

$$f(u) = \gamma$$

where $u(x) = \gamma x + a$. It is clear that f is a surjective homomorphism from $\Gamma(E, H)$ onto the multiplicative group $\mathbb{K}^*$, and we have

$$\Theta(E, H) = \ker f;$$

consequently we have

$$\frac{\Gamma(E, H)}{\Theta(E, H)} \cong \mathbb{K}^*.$$

Proposition 3.16 *Let $\Theta'(E, T)$ be the set of all transvections with a given line T and with variable hyperplane H containing T. Then $\Theta'(E, T)$ is an abelian group.*

Proposition 3.17 *Let u, u' be respectively two non-identity dilations or transvections with hyperplanes H, H', of lines D, D' respectively. We have:*

1. *A necessary and sufficient condition for u to commute with u' is that either $H' = H$ or $D \subset H'$ and $D' \subset H$. This condition is also sufficient if u, u' are both transvections.*

2. *If u, u' are both dilations, the necessary and sufficient condition for commutativity is that either $H = H'$ and $D = D'$ or $D \subset H'$ and $D' \subset H$.*

3. *A dilation u never commutes with a transvection u' unless one or both is the identity map.*

Proposition 3.18 *Let $\tau(x) = x + \varphi(x)t$ be a transvection with hyperplane H and line $\mathbb{K}t$. The following properties are equivalent:*

1. $\tau \neq \mathrm{id}_E$,

2. $t \neq 0$ and $\varphi \neq 0$.

This is an immediate consequence of the characterization of transvections.

3.5.2 k-symplectic transvections

Let E be an $n(k+1)$-dimensional vector space over a commutative field $\mathbb{K}$ equipped with a k-symplectic exterior system $(\theta^1, \ldots, \theta^k; F)$.

Definition 3.11 *A transvection τ of E with fixed hyperplane H and line $\mathbb{K}t$ is called k-symplectic if it belongs to the k-symplectic group $Sp(k, n; E)$.*

It follows that a transvection $\tau(x) = x + \varphi(x)t$ with fixed hyperplane H and line $\mathbb{K}t$ is k-symplectic if and only if, the following property is satisfied

$$\varphi(x)\theta^p(y, t) = \varphi(y)\theta^p(x, t) \tag{3.4}$$

for all $x, y \in E$ and $p(p = 1, \ldots, k)$.

Proposition 3.19 *If $\tau(x) = x + \varphi(x)t$ is a k-symplectic transvection with fixed hyperplane H and line $\mathbb{K}t$ such that $\tau \neq \mathrm{id}_E$, then $H = t^\perp$.*

Proof. The relation 3.4 implies that $H \subseteq t^\perp$, whilst the inclusion $t^\perp \subseteq H$ results from the non-degeneracy of the system $\{\theta^1, \cdots, \theta^k\}$ and from the relationship 3.4. ∎

Proposition 3.20 *Let* $\tau(x) = x + \varphi(x)t$ *be a k-symplectic transvection with fixed hyperplane* H *and line* $\mathbb{K}t$, *such that* $\tau \neq \mathrm{id}_E$. *Then for each* p $(p = 1, \ldots, k)$, *the following properties are equivalent:*

1. *There exists* $x \in E - H$ *such that* $\theta^p(t, x) \neq 0$;

2. *For every* $x \in E - H$, $\theta^p(t, x) \neq 0$.

This proposition is an immediate consequence of 3.4.

For each element $t \in E - \{0\}$ we associate the set $I(t)$ of elements $p \in \{1, \ldots, k\}$ such that $i(t)\theta^p \neq 0$. The non-degeneracy of the system $\{\theta^1, \ldots, \theta^k\}$ implies that $I(t)$ is non empty. Let $\tau(x) = x + \varphi(x)t$ be a k-symplectic transvection with fixed hyperplane H and line $\mathbb{K}t$; the relationship 3.4 and the preceding proposition show that we have

$$\frac{\varphi(x)}{\varphi(y)} = \frac{\theta^p(t, x)}{\theta^p(t, y)}$$

for all $x, y \in E - H$ and $p \in I(t)$. Then there exists non zero scalars μ^p $(p \in I(t))$ such that

$$\varphi(x) = \mu^p \theta^p(x, t)$$

for all $x \in E - H$ and $p \in I(t)$. This equation is independent of p in $I(t)$, thus we have

$$\mu^p \theta^p = \mu^q \theta^q$$

for all $p, q \in I(t)$.

The following proposition allows us to give a characterization of k-symplectic transvections.

Proposition 3.21 *Let* t *be a vector of* E. *A necessary and sufficient condition for* t *to be a direction of a* k-*symplectic transvection* τ *with* $\tau \neq \mathrm{id}_E$ *is that,* t *takes the form*

$$t = \sum_{p=1}^{k} \left(\sum_{i=1}^{n} t^i \lambda^p e_{pi} \right), \tag{3.5}$$

where t^i *(resp.* λ^p*) are non-zero scalars.*

Proof. For $k = 1$ it suffices to show that $t \in F$. In fact, we have $t^{\perp} = H = \ker \varphi$, and F is τ-invariant. Thus $t \in F$.

Suppose that $k \geq 2$. Let $\tau(x) = x + \varphi(x)t$ be a k–symplectic transvection with hyperplane H and line $\mathbb{K}t$, with $\tau \neq \mathrm{id}_E$. The previous relationships prove that there exists a subset $I(t)$ of $\{1, \ldots, k\}$ such that the linear forms $i(t)\theta^p$, $p \in I(t)$, are proportional. We can assume that $1 \in I(t)$. Write

$$t = \sum_{p=1}^{k} \left(\sum_{i=1}^{n} t^{pi} e_{pi} \right) + \sum_{i=1}^{n} t^i e_i.$$

If $I(t)$ is reduced to $\{1\}$, then $i(t)\theta^p = 0$ for every $p \geq 2$, and $t = \sum_{i=1}^{n} t^{1i} e_{1i}$. This proves that t can be written in the form 3.5. Suppose now that $I(t)$ is not reduced to $\{1\}$. For every $q \in I(t)$ there exists a non-zero scalar λ^q such that $i(t)\theta^q = \lambda^q i(t)\theta^1$; thus

$$t^{qi} = \lambda^q t^{1i}, \quad t^i = 0$$

for every $i \in \{1, \ldots, n\}$, and t is written in the expected form.

Conversely, let t be an element of $E - \{0\}$ of the form 3.5. It is clear that $I(t)$ is the set of $q \in \{1, \ldots, k\}$ such that $\lambda^q \neq 0$ and

$$\sum_{i=1}^{n} t^i \omega^i = 0.$$

is an equation of the hyperplane H, and we have

$$i(t)\theta^q = \lambda^q \left(\sum_{i=1}^{n} t^i \omega^i \right) \tag{3.6}$$

for every $q \in I(t)$, then

$$H = \ker(i(t)\theta^1) \cap \cdots \cap \ker(i(t)\theta^k) = t^{\perp}.$$

The relationship 3.6 implies that there exist non zero elements c^q ($q \in I(t)$) of $\mathbb{K}$, and a linear form φ of which kernel is H, such that

$$c^q i(t)\theta^q = \varphi$$

for every $q \in I(t)$. The endomorphism τ of E given by:

$$\tau(x) = x + \varphi(x)t,$$

for every $x \in E$, is a k–symplectic transvection of E. ∎

Corollary 3.8 *For every k-symplectic transvection of E with fixed hyperplane H we have $F \subseteq H$.*

Corollary 3.9 *For a transvection τ of E the following properties are equivalent:*

1. τ *is a k-symplectic transvection;*

2. τ *is of the form*

$$\tau(x) = x + \left(\sum_{m=1}^{n} t^m \omega^m(x) \right) \left(\sum_{m=1}^{n} \sum_{p=1}^{k} t^m \lambda^p e_{pm} \right).$$

3.5.3 The group $Tp(k, n; E)$

We denote by $Tp(k, n; E)$ the subgroup of $Sp(k, n; E)$ spanned by the k-symplectic transvections of E. As in preceding sections, we will denote by $Tp(k, n; \mathbb{K})$ the group of matrices of elements of $Tp(k, n; E)$ relative to the k-symplectic basis.

Proposition 3.22 *The group $Tp(k, n; \mathbb{K})$ is the set of the matrices:*

$$\begin{pmatrix} I_n & \cdots & 0 & S_1 \\ \vdots & \ddots & & \vdots \\ 0 & \cdots & I_n & S_k \\ 0 & \cdots & 0 & I_n \end{pmatrix}, \tag{3.7}$$

where I_n is the unit $n \times n$ matrix and $S_1, \cdots, S_k$ are $n \times n$ symmetric matrices with coefficients in $\mathbb{K}$. In particular, $Tp(k, n; \mathbb{K})$ is a normal abelian subgroup of $Sp(k, n; \mathbb{K})$.

Proof. It follows from the previous corollary that for every k-symplectic transvection we have:
$$\tau(x) = x$$
for each $x \in F$. Thus every element of $Tp(k, n; \mathbb{K})$ is of the form 3.7.

Conversely, let M be a matrix of the type 3.7. Consider the endomorphisms τ_r^{pij} and ρ_r^{pi} of E defined by:

$$\begin{aligned} \tau_r^{pij}(x) &= x + \left(r\omega^i(x) + \omega^j(x) \right) \left(r e_{pi} + e_{pj} \right), \\ \rho_r^{pi}(x) &= x + r\omega^i(x) e_{pi}. \end{aligned}$$

where $p = 1, \ldots, k$, $i, j = 1, \ldots, n$ and $r \in \mathbb{K}$. The mappings τ_r^{pij} and ρ_r^{pi} are k-symplectic transvections, contrarly to the composition

$$\sigma_r^{pij} = \tau_r^{pij} \circ (\rho_{r^2}^{pi})^{-1} \circ (\rho_1^{pj})^{-1},$$

and we verify the following relationships:

1. $\tau_r^{pii} = \rho_{(1+r)^2}^{pi}$;

2. $\sigma_r^{pij}(x) = x + r\left(\omega^i(x)e_{pj} + \omega^j(x)e_{pi}\right)$;

3. $\sigma_r^{pij}(e_i) = e_i + re_{pj}$;

4. $\sigma_r^{pij}(e_j) = e_j + re_{pi}$;

5. $\sigma_r^{pij}(e_{i'}) = e_{i'}$ if $i' \neq j$ and $i' \neq j$.

For every p $(p = 1, \ldots, k)$ consider the product

$$\sigma_p = \left(\prod_{1 \leq i \leq n} \rho_{s_{(p)}^{ii}}^{pi} \right) \left(\prod_{1 \leq i < j \leq n} \sigma_{s_{(p)}^{ij}}^{pij} \right),$$

where s_p^{ij} $(i, j = 1, \ldots, n)$ are the coefficients of the matrix S_p for any p $(p = 1, \ldots, k)$. With respect to the k-symplectic basis of E the matrix of σ_p has the form (3.7), where $S_q = 0$ for $q \neq p$. It results that M is the matrix of compositions $\sigma_1 \circ \cdots \circ \sigma_k$ with respect to this basis. ∎

3.5.4 The affine group $Hp(k, n; E)$

In this subsection we assume that $\mathbb{K} = \mathbb{R}$ or $\mathbb{C}$. The group of all affine transformations

$$X \longmapsto AX + B$$

of $\mathbb{K}^{n(k+1)}$ with $A \in Tp(k, n; \mathbb{K})$ is a Lie group denoted by $Hp(k, n; \mathbb{K})$. The elements of this group are matrices

$$\begin{pmatrix} I_n & 0 & \cdots & 0 & S_1 & T_1 \\ \cdots & \cdots & \cdots & \cdots & \cdots & \cdots \\ 0 & \cdots & \cdots & I_n & S_k & T_k \\ 0 & \cdots & \cdots & 0 & I_n & Q \\ 0 & \cdots & \cdots & 0 & 0 & 1 \end{pmatrix}$$

where I_n is the unit $n \times n$ matrix, $S_1, \cdots,\ S_k$ are $n \times n$ symmetric matrices with coefficients in $\mathbb{K}$, and $T_1, \ldots, T_k$, Q are column matrices.

We denote by

$$(S_1, \ldots, S_k, Q, T_1, \ldots, T_k)$$

the elements of this group. For all $g = (S_1, \ldots, S_k, Q, T_1, \ldots, T_k)$ and $g' = (S'_1, \ldots, S'_k, Q', T'_1, \ldots, T'_k)$ belonging to $Hp(k, n; \mathbb{K})$, we have:

1. $gg' = (S_1 + S'_1, \ldots, S_k + S'_k, Q + Q', S_1 Q' + T_1 + T'_1, \ldots,$
   $$S_k Q' + T_k + T'_k),$$

2. $g^{-1} = (-S_1, \ldots, -S_k, -Q, S_1 Q - T_1, \ldots, S_k Q - T_k),$

3. $[g, g'] = (0, \ldots, 0, 0, S_1 Q' - S'_1 Q, \ldots, S_k Q' - S'_k Q).$

Proposition 3.23 *The group $Hp(k, n; \mathbb{K})$, is a nilpotent connected and simply connected Lie group such that $[Tp(k, n; \mathbb{K}), Tp(k, n; \mathbb{K})]$ coincides with the center of $Tp(k, n; \mathbb{K})$.*

Chapter 4

PFAFFIAN SYSTEMS

4.1 Differential exterior systems

A good introduction into this field is given in [13]. We present here only the base of the theory of exterior systems and Pfaffian systems. For a deep study of exterior systems we refer the reader to this reference [13].

4.1.1 Differential p-forms

Let $\mathbb{R}^n$ be the real space of dimension n equipped with a coordinates system $(x_1, \cdots, x_n)$, let $T\mathbb{R}^n$ be the tangent bundle of $\mathbb{R}^n$, and let $T^*\mathbb{R}^n$ be its cotangent bundle. We denote by $\Lambda^p T^*\mathbb{R}^n$ the vector bundle over $\mathbb{R}^n$ whose fibres are $\Lambda^p T_x^*\mathbb{R}^n$, $x \in \mathbb{R}^n$ (the space of skew symmetric p-forms of the real space $T_x\mathbb{R}^n$).

Definition 4.1 *An exterior differential form of degree p (or simply a differential p-form) on $\mathbb{R}^n$ is a smooth section of the vector bundle*

$$\Lambda^p T^*\mathbb{R}^n \longrightarrow \mathbb{R}^n.$$

A Pfaffian form is an exterior differential form of degree 1.

With respect to the coordinate system $(x_1, \cdots, x_n)$ the derivatives

$$(\partial/\partial x_1(x), \cdots, \partial/\partial x_n(x))$$

is a basis of $T_x\mathbb{R}^n$,

$$(dx_1(x), \cdots, dx_n(x))$$

is a basis of $T_x^* \mathbb{R}^n$ and any exterior differential p-form α has the form:

$$\alpha = \sum \alpha_{i_1 \cdots i_p} dx_{i_1} \wedge \cdots \wedge dx_{i_p},$$

where $\alpha_{i_1 \cdots i_p}$ are smooth functions with respect to the coordinates system $(x_1, \cdots, x_n)$, and $\alpha(x)$ is an exterior p-form of the tangent space $T_x \mathbb{R}^n$. By definition we have:

1. a differential 0-form is a scalar function on $\mathbb{R}^n$;

2. each differential p-form α is zero ($\alpha = 0$) for $p > n$;

3. any differential n-form α can be written as

$$\alpha = f(x) dx_1 \wedge \cdots \wedge dx_n,$$

where f is a smooth function with respect to the coordinates system $(x_1, \cdots, x_n)$.

4.1.2 Exterior algebra

The set of exterior differential p-forms on $\mathbb{R}^n$ is a real vector space. The sum and the multiplication by a scalar are defined as follows:let α and β be two differential p-forms given by:

$$\alpha = \sum \alpha_{i_1 \cdots i_p} dx_{i_1} \wedge \cdots \wedge dx_{i_p},$$

$$\beta = \sum \beta_{i_1 \cdots i_p} dx_{i_1} \wedge \cdots \wedge dx_{i_p},$$

then $\alpha + \beta$ is the p-form

$$\alpha + \beta = \sum (\alpha_{i_1 \cdots i_p} + \beta_{i_1 \cdots i_p}) dx_{i_1} \wedge \cdots \wedge dx_{i_p}$$

and for every $\lambda \in \mathbb{R}$ the p-form $\lambda \alpha$ is given by

$$\lambda \alpha = \sum \lambda \alpha_{i_1 \cdots i_p} dx_{i_1} \wedge \cdots \wedge dx_{i_p}.$$

For every smooth function f on $\mathbb{R}^n$ the product of f by the p-form α is defined by

$$f \alpha = \sum f \alpha_{i_1 \cdots i_p} dx_{i_1} \wedge \cdots \wedge dx_{i_p}.$$

Therefore, the set of all exterior differential p-forms is a module over the ring of smooth functions on $\mathbb{R}^n$.

Let $\alpha \in \Lambda^p(\mathbb{R}^n)$ and $\beta \in \Lambda^q(\mathbb{R}^n)$. Define $\alpha \wedge \beta \in \Lambda^{p+q}(\mathbb{R}^n)$ by

$$(\alpha \wedge \beta)(x) = \alpha(x) \wedge \beta(x)$$

for every $x \in \mathbb{R}^n$. Note that we have:

$$f(\alpha \wedge \beta) = f\alpha \wedge \beta = \alpha \wedge f\beta.$$

The space of exterior differential forms equipped with the exterior product $\wedge$ is an associative algebra.

4.1.3 Pull back of an exterior form by a mapping

Let $f : \mathbb{R}^n \longrightarrow \mathbb{R}^m$ be a smooth mapping and let α be an exterior differential p-form on $\mathbb{R}^m$. The pull back of α by f is the differential p-form $f^*\alpha$ defined on $\mathbb{R}^n$ by

$$(f^*\alpha)(x) = \sum \alpha_{i_1 \cdots i_p}(y(x)) \frac{\partial y_{i_1}}{\partial x_{j_1}} \frac{\partial y_{i_2}}{\partial x_{j_2}} \cdots \frac{\partial y_{i_p}}{\partial x_{j_p}} dx_{j_1} \wedge \cdots \wedge dx_{i_{jp}}$$

where $\alpha(y) = \sum \alpha_{i_1 \cdots i_p}(y) dy_{i_1} \wedge \cdots \wedge dy_{i_p}$ and $y = f(x)$. We deduce that

$$f^*(\alpha \wedge \beta) = f^*\alpha \wedge f^*\beta$$

for all exterior differential p-forms α and β on $\mathbb{R}^m$.

4.1.4 Exterior derivatives of p-forms

First, we recall the definition of the exterior derivative of 0-form. Let $f : \mathbb{R}^n \longrightarrow \mathbb{R}$ be a smooth function. In the coordinates system $(x_1, \ldots, x_n)$, the exterior derivative df is given by:

$$df(x) = \sum \frac{\partial f}{\partial x_i}(x) dx_i.$$

We generalize this to general p-forms as follows: the exterior derivative of a p-form

$$\alpha = \sum \alpha_{i_1 \cdots i_p} dx_{i_1} \wedge \cdots \wedge dx_{i_p}$$

is

$$d\alpha = \sum d\alpha_{i_1 \cdots i_p} \wedge dx_{i_1} \wedge \cdots \wedge dx_{i_p},$$

where, of course,

$$d\alpha_{i_1 \cdots i_p} = \sum \frac{\partial \alpha_{i_1 \cdots i_p}}{\partial x_i}(x) dx_i.$$

This operator verifies the following properties :

1. $d(\alpha + \beta) = d\alpha + d\beta$;

2. $d(\alpha \wedge \beta) = d\alpha \wedge \beta + (-1)^p \alpha \wedge d\beta$ if α is an exterior differential p-form

3. $d \circ d = 0$;

4. $d(f^*\alpha) = f^* d\alpha$.

4.1.5   Exterior differential systems

Definition 4.2 *A differential exterior system on $\mathbb{R}^n$ is a system of equations*

$$\theta_1 = 0, \cdots, \theta_s = 0,$$

where $\theta_1, \cdots, \theta_s$ are exterior differential forms.

Example 8 *Considering a first order partial differential equations*

$$F\left(x_1, \cdots, x_n, y, \frac{\partial y}{\partial x_1}, \cdots, \frac{\partial y}{\partial x_n}\right) = 0.$$

Taking $p_i = \partial y / \partial x_i$, we obtain the following exterior differential system:

$$\begin{cases} F(x_1, \cdots, x_n, y, p_1, \cdots, p_n) = 0, \\ \\ dy - \sum p_i dx_i = 0 \end{cases}$$

in the real space $\mathbb{R}^{2n+1}$ parametrized by $(x_1, \cdots, x_n, y, p_1, \cdots, p_n)$.

Definition 4.3 *Let (S) be an exterior differential system on $\mathbb{R}^n$. An integral manifold of (S) is a submanifold W of $\mathbb{R}^n$ such that for every $x \in W$ the tangent space $T_x W$ is a solution of the exterior system defined by $\{\theta_1(x) = 0, \cdots, \theta_s(x) = 0\}$.*

If $f : \mathbb{R}^n \longrightarrow \mathbb{R}^n$ is a mapping such that W corresponds to $f(x) = 0$, then W is an integral manifold of (S) if and only if $f^*\theta_i = 0$ for each $i = 1, \cdots, s$. Let $(S) = (\theta_1 = 0, \cdots, \theta_s = 0)$ be an exterior differential system on $\mathbb{R}^n$.

Definition 4.4 *The exterior differential system $F(S)$ defined by the equations $\theta_1 = 0, \cdots, \theta_s = 0, d\theta_1 = 0, \cdots, d\theta_s = 0$ is called the closure of (S).*

Proposition 4.1 *An exterior differential system (S) and its closure $F(S)$ have the same integral manifolds.*

Proof. In fact, let W be the submanifold of $\mathbb{R}^n$ defined by $f(x) = 0$. Then we have $f^*\theta_i = 0$ for each $i = 1, ..., s$. Since $d(f^*\theta_i) = f^*d\theta_i$, therefore $f^*d\theta_i = 0$, which proves that W is an integral manifold of $(F(S))$. ∎

Definition 4.5 *The exterior differential system (S) is closed if it is algebraically equivalent to $(F(S))$, that is, if it generates the same ideal of $\Lambda(\mathbb{R}^n)$ (algebra of exterior differential forms on $\mathbb{R}^n$).*

4.2 Pfaffian systems

4.2.1 Definition

Definition 4.6 *A Pfaffian system is an exterior differential system*

$$(\alpha_1 = 0, \cdots, \alpha_s = 0),$$

where α_i are Pfaffian forms, that is, exterior differential forms of degree 1.

An integral manifold of such a system is a submanifold W such that the tangent space T_xW at each point $x \in W$ is contained in the kernel of the linear system $(\alpha_1(x) = 0, \cdots, \alpha_s(x) = 0)$.

4.2.2 Algebraically equivalent Pfaffian systems

Let $(\alpha_1 = 0, \cdots, \alpha_s = 0)$ and $(\beta_1 = 0, \cdots, \beta_t = 0)$ be two Pfaffian systems on $\mathbb{R}^n$ we say that they are algebraically equivalent if there exists smooth functions g_i^j and h_j^i, such that

$$\beta_i = \sum g_i^j \alpha_j$$

and

$$\alpha_j = \sum h_j^i \beta_i.$$

For example, if the Pfaffian forms α_i are independent (that is, the linear forms $\alpha_i(x)$ are independent at each point $x \in \mathbb{R}^n$) and the matrix $(g_i^j(x))$ is invertible at each point x, then the systems $(\alpha_1 = 0, \cdots, \alpha_s = 0)$ and $(\beta_1 = 0, \cdots, \beta_s = 0)$ are algebraically equivalent, where $\beta_i = \sum g_i^j \alpha_j$.

Two algebraically equivalent Pfaffian systems are the same integral manifolds. This follows from the previous definition and the study of algebraically equivalent systems.

The main purpose of the two following chapters is to consider the problem of the classification of Pfaffian systems, which brings upback to seeing how to represent locally all Pfaffian system on $\mathbb{R}^n$. This necessitates a study of the invariants of Pfaffian systems, that is, those objects depending only on the class of algebraically independent systems.

Remark 5 *Let $(S) = (\alpha_1 = 0, \cdots, \alpha_r = 0)$ be a Pfaffian system. To simplify, one will write $(S) = (\alpha_1, \cdots, \alpha_r)$ for $\alpha_1 = 0, \cdots, \alpha_r = 0$. Similarly, if $\beta = \sum a_i \alpha_i$, then $\beta = 0$ is an equation of the system, and we will write $\beta \in (S)$.*

4.3 Class of Pfaffian systems

4.3.1 Rank of Pfaffian systems

Let (S) be a Pfaffian system on $\mathbb{R}^n$. The rank of (S) at a point x is the rank of the linear system $(\alpha_1(x), \alpha_2(x), \cdots, \alpha_r(x))$. We denote by $\text{rank}_x(S)$ the rank of (S). Generally, the mapping $x \longmapsto \text{rank}_x(S)$ is not constant; it is upper semi-continuous only. In this work we consider only those Pfaffian systems of constant rank.

This hypothesis avoids the study of the singularity and to make a study of Pfaffian systems in a neighborhood of a generic point, where the rank is maximum. The aim is a local study of Pfaffian systems, and this convention is not too restrictive.

4.3.2 Class of Pfaffian systems

Let (S) be a Pfaffian system of rank r on $\mathbb{R}^n$ represented by the Pfaffian forms $(\alpha_1, ..., \alpha_r)$. *The characteristic space of (S) at a point $x \in \mathbb{R}^n$ is the space $C_x(S)$ defined by:*

$$\{X \in T_x\mathbb{R}^n \mid \alpha_i(x)(X) = 0, X \lrcorner (d\alpha_i \wedge \alpha_1 \wedge ... \wedge \alpha_r)_x = 0, \ \forall i = 1, ..., r \}$$

Recall that $X \lrcorner \alpha$ (noted also $i(X)\alpha$) is the interior product of the exterior form α with respect to the vector X.

Definition 4.7 *The class of (S) at a point x (denoted by $\text{class}_x(S)$) is the codimension of the vector space $C_x(S)$.*

This definition equally admits a dual presentation. One considers the closure $(F(S))$ of the Pfaffian system (S), that is, the following exterior differential system:

$$(F(S)) = (\alpha_i = 0,\ d\alpha_i = 0,\ i = 1, ..., r).$$

The associated Pfaffian system which corresponds at each point to the linear system associated to $F_x(S)$ is defined by

$$A_x^* (F(S)) = \{\alpha_i (x) = 0,\ X_j (x) \rfloor d\alpha_i (x) = 0,\ i = 1, ..., r,\ j = 1, ..., n\}$$

where $(X_1, ..., X_n)$ is a local basis of vector fields on $\mathbb{R}^n$. The rank of $A_x^* (F(S))$ corresponds to the class of (S) at the point x.

Definition 4.8 *A characteristic vector field of (S) is a vector field contained in the distribution $C(S)$.*

4.3.3 Interpretation of the class

Let $(S) = (\alpha_1 = 0, ..., \alpha_r = 0)$ be a Pfaffian system of constant rank r on $\mathbb{R}^n$. The kernel of (S) at a point $x \in \mathbb{R}^n$ is given by

$$\ker_x (S) = \{X_x \in T_x\mathbb{R}^n \mid \alpha_i (x) (X_x) = 0,\ i = 1, ..., r\}.$$

It determines a distribution on $\mathbb{R}^n$ of codimension r. We have defined the notion of integrable system. The original problem of Pfaffian systems was concerned with research on integral manifolds. The class of a Pfaffian system is an invariant measuring the non-integrability of this system. It is an invariant because this number is independent of the choice of representatives $\alpha_1, ..., \alpha_r$ of (S). Two systems algebraically equivalent have the same class at each point. Of course, the class of (S) is a function whose value depends on the point x where it is evaluated. It is upper semi-continuous superiorly.

Therefore we can study the Pfaffian system on a neighborhood of a point where the class is a maximum. This brings us to making the next hypothesis: all Pfaffian systems considered in this work are assumed to be of constant class.

4.4 Integrability of Pfaffian systems

4.4.1 Completely integrable Pfaffian systems

Definition 4.9 *A Pfaffian system* (S) *of rank* r *on* $\mathbb{R}^n$ *is called completely integrable if for each* $p \in \mathbb{R}^n$ *there exists a coordinate system* $(y_1, ..., y_n)$ *defined on an open neighborhood of* p, *called a distinguished coordinate system, such that* (S) *is defined by the Pfaffian system*

$$(dy_1 = 0, ..., dy_r = 0).$$

If (S) is completely integrable the mapping

$$x \longmapsto (y_1(x), ..., y_r(x))$$

is of constant rank r. For every point $(a_1, \ldots, a_n) \in \mathbb{R}^n$ this mapping defines an integral manifold passing through the point whose equations are $y_i(x) = a_i$. The functions y_i are called *first integrals* of the system.

Proposition 4.2 *Each completely integrable Pfaffian system* (S) *is of constant class equal to its rank.*

Proof. In fact,

$$C_x(S) = \{X_x \in T_x\mathbb{R}^n \mid \alpha_i(x)(X_x) = 0 \text{ and } X_x \rfloor d\alpha_i(x) = 0 \mod (S)\}.$$

The notation $X_x \rfloor d\alpha_i(x) = 0 \mod (S)$ means that we have

$$X_x \rfloor d\alpha_i(x) \wedge \alpha_1(x) \wedge \ldots \wedge \alpha_r(x) = 0.$$

We can choose as representative for (S) the Pfaffian forms $\alpha_i = dy_i$. We then deduce that we have $d\alpha_i = 0$, and

$$C_x(S) = \{X_x \in T_x\mathbb{R}^n \mid \alpha_i(x)(X_x) = 0\} = \ker_x(S).$$

Thus we have $\dim C_x(S) = n - r$, and $\operatorname{codim} C_x(S) = \operatorname{rank}_x(S) = r$. ∎

4.4.2 Frobenius' theorem

Theorem 4.1 *Let* (S) *be a Pfaffian system of rank* r *on* $\mathbb{R}^n$. *Then* (S) *is completely integrable, if and only if, the class of* (S) *is constant and it coincides with the rank of this system* $(\operatorname{rank}_x(S) = \operatorname{class}_x(S))$.

Proof. We prove this theorem by induction on the dimension of $\ker(S)$.

1. If (S) is of rank $n-1$, that is, $\dim \ker(S) = 1$, then for every $p \in \mathbb{R}^n$ we can find a coordinates system $(y_1, ..., y_n)$ defined on an open set U of $\mathbb{R}^n$ containing the point p such that $\left\{ \frac{\partial}{\partial y_1} \right\}$ is a basis of $\ker(S)$. Consequently the system (S) is spanned by $(dy_2, ..., dy_n)$, then it is completely integrable. Note here that the hypothesis relating to the class is automatically verified:if $\mathrm{rank}(S) = n-1$, then $\mathrm{class}(S) = \mathrm{rank}(S) = n-1$.

2. Assume that all Pfaffian system of a given rank $r+1$ of which class coincides with the rank is completely integrable, and consider a system (S) of rank r. Let $(\alpha_1, ..., \alpha_r)$ be a basis of (S). With respect to the coordinates system (x_i) defined on an open subset U centered at the origin O the Pfaffian forms α_i are written:

$$\left\{ \begin{array}{l} \alpha_1 = a_{11}dx_1 + ... + a_{1n}dx_n \\ \vdots \\ \alpha_r = a_{r1}dx_1 + ... + a_{rn}dx_n. \end{array} \right.$$

Since the rank of (S) is equal to r the matrix (a_{ij}) is of rank r. We can assume that the matrix $\widetilde{M} = (a_{ij})_{1 \leq i,j \leq r}$ is invertible at each point of U. The Pfaffian forms given by

$$(\beta_i) = \widetilde{M}^{-1}(\alpha_i)$$

satisfy

$$\left\{ \begin{array}{l} \beta_1 = dx_1 + \sum_{j>r} b_{ij}dx_j \\ \vdots \\ \beta_r = dx_r + \sum_{j>r} b_{rj}dx_j \end{array} \right.$$

and also span (S). We deduce that the forms

$$(\beta_1, ..., \beta_r, dx_{r+1}, ..., dx_n)$$

are linearly independent on U. Since $\mathrm{class}(S) = r$, the class of the Pfaffian system $\left(\widetilde{S} \right) = (\beta_1, ..., \beta_r, dx_{r+1})$ is equal to its rank

$$\mathrm{class} \left(\widetilde{S} \right) = \mathrm{rank} \left(\widetilde{S} \right) = r+1.$$

By hypothesis there exists a coordinates system $(y_1,...,y_n)$ such that $\left(\widetilde{S}\right)$ is spanned by $(dy_1,...,dy_{r+1})$. Thus

$$dx_{r+1} = \sum_{j=1}^{r+1} c_j dy_j.$$

We can assume that $c_{r+1} \neq 0$. Also the system $(\widetilde{S})$ is spanned by the forms $(dy_1,...,dy_r,dx_{r+1})$ thus the system (S) is generated by a system of the following form:

$$\begin{cases} \widetilde{\beta}_1 = dy_1 + a_1 dx_{r+1}, \\ \qquad \vdots \\ \widetilde{\beta}_r = dy_r + a_r dx_{r+1}. \end{cases}$$

We deduce that

$$\begin{cases} \widetilde{d\beta}_1 = da_1 \wedge dx_{r+1}, \\ \qquad \vdots \\ \widetilde{d\beta}_r = da_r \wedge dx_{r+1}. \end{cases}$$

By hypothesis the class of the system $(\widetilde{\beta}_1,...,\widetilde{\beta}_r)$ is equal to r. Thus the forms $\widetilde{\beta}_i$ and $d\widetilde{\beta}_i$ are expressed only by r differentials, and are in the ideal spanned by these r differentials. Therefore we have

$$da_i = \sum_{j=1}^{r} \frac{\partial a_i}{\partial y_j} dy_j + \frac{\partial a_i}{\partial x_{r+1}} dx_{r+1},$$

$(i = 1,...,r)$. Hence

$$d\widetilde{\beta}_i = \sum_{j=1}^{r} \frac{\partial a_i}{\partial y_j} dy_j \wedge dx_{r+1} \qquad (i = 1,...,r).$$

On the open $\widetilde{U}$ equipped with the coordinates system $(y_1,...,y_r,x_{r+1})$, the Pfaffian system $\left(\widetilde{\beta}_{1i},...,\widetilde{\beta}_r\right)$ is of rank equal to dim $\widetilde{U}$ -1. By induction there exists local coordinates $(z_1,...,z_r,z_{r+1})$ such that the forms

$$\widetilde{\beta}'_1 = dz_1,...,\widetilde{\beta}'_r = dz_r$$

generate this system, that is it is algebraically equivalent to (S). ■

4.4.3 Characterization of completely integrable system

Theorem 4.2 *A Pfaffian system* (S) *is completely integrable if and only if for any representative* $(\alpha_1, ..., \alpha_r)$ *we have*

$$d\alpha_i \wedge \alpha_1 \wedge ... \wedge \alpha_r = 0,$$

for all $i = 1, ..., r$, *that is, if* (S) *is closed.*

Proof. Assume that (S) is completely integrable. Then

$$\text{class}\,(S) = \text{rank}\,(S) = r.$$

Thus rank $(A^*(F(S))) = r$ at each point. By definition, $A^*(F(S))$ contains (S), thus each set of generators $(\alpha_1, ..., \alpha_r)$ of (S) generates the closure of (S). The differentials $d\alpha_i$ belong to the ideal spanned by $(\alpha_1, ..., \alpha_r)$, this means that we have

$$d\alpha_i \wedge \alpha_1 \wedge ... \wedge \alpha_r = 0$$

for every $i = 1, ..., r$.

Conversely, if this last relationship is satisfied for each set of generators of (S), the class of (S) is equal to its rank and (S) is completely integrable. ∎

Such a theorem admits a dual version.

Theorem 4.3 *Let* (S) *be a Pfaffian system of rank* r. *Then* (S) *is completely integrable if and only if, there exists a basis* $(X_1, ..., X_{n-r})$ *of the kernel* ker (S) *satisfying*

$$[X_i, X_j] = \sum_{k=1}^{n-r} C_{ij}^k X_k.$$

Proof. Let $(\alpha_1, ..., \alpha_r)$ be generators of (S) which we extend to a system $(\alpha_1, ..., \alpha_r, \beta_{r+1}, ..., \beta_n)$ of n forms linearly independent at each point. Since (S) is completely integrable we have

$$d\alpha_i = \sum C_{jk}^i \alpha_j \wedge \alpha_k + \sum \tilde{C}_{jk}^i \alpha_j \wedge \beta_k.$$

If $(Y_1, ..., Y_r, X_{r+1}, ..., X_n)$ is the dual basis, the vector fields $X_{r+1}, ..., X_n$ form a basis of the kernel of (S). The duality relations between the brackets and exterior differentials imply that we have

$$[X_i, X_j] = \sum_{k=r+1}^{n} C_{ij}^k X_k,$$

for all $i, j = r + 1, ..., n$. This proves the theorem. ∎

4.5   The Darboux theorem

We have seen above that all completely integrable Pfaffian systems of rank r on $\mathbb{R}^n$ (that is, of class r) are locally isomorphic. It is similarly so for all systems of rank 1 and maximal class (or, more generally, constant class).

4.5.1   Class of Pfaffian systems of rank 1.

Considering a Pfaffian system (S) of rank 1 defined by the Pfaffian equation $\alpha = 0$.

Proposition 4.3 *The class of* $(S) = (\alpha)$ *at a point* x *is equal to* $2s + 1$ *if and only if*

$$(\alpha \wedge (d\alpha)^s)_x \neq 0 \quad and \quad \left(\alpha \wedge (d\alpha)^{s+1}\right)_x = 0. \tag{4.1}$$

Proof.   Assume that the rank at x of the 2-form $d\alpha$ is $2r$. Then we have

$$(d\alpha)^r \neq 0 \text{ and } (d\alpha)^{r+1} = 0.$$

Two cases are possible:

1. $\alpha \wedge (d\alpha)^r \neq 0$ and $\alpha \wedge (d\alpha)^{r+1} = 0$;

2. $\alpha \wedge (d\alpha)^r = 0$ and $\alpha \wedge (d\alpha)^{r-1} \neq 0$.

In the first case we obtain $\dim C_x(S) = n - 2r - 1$, and in the second case we have $\dim C_x(S) = n - 2r + 1$ which implies that class $\{\alpha\} = 2r + 1$ or $2r - 1$. The reverse is immediate. ∎

Remark 6 *This proposition gives a very practical criterion determining the class of a Pfaffian system of rank 1. Note that we have no generalization of this nature for systems whose rank is greater than 2. Also we can observe here that the class represents the number of independent differentials permitting the 2-form $d\alpha$ and the form α to be writting.*

4.5.2   The Darboux theorem

Theorem 4.4 *Let* $(S) = \{\alpha = 0\}$ *be a Pfaffian system of rank 1 and of constant class* $2s + 1$ *in* $\mathbb{R}^n$. *Then there exists a local coordinate system*

$$(x_1, ..., x_s, y_1, ..., y_s, z, u_1, ..., u_{n-2s-1}),$$

such that

$$\alpha = dz + x_1 dy_1 + ... + x_s dy_s.$$

Proof.   First time we assume that n is odd and that $(S) = \{\alpha\}$ is of maximum class n, that is,

$$\alpha \wedge (d\alpha)^s \neq 0 \text{ with } n = 2s + 1.$$

Lemma 9 *Suppose that $\{\alpha\}$ is of maximum class. For every integrable distribution D contained in $\ker \alpha$, we have:*

$$\dim D \leq s.$$

Proof of lemma.   Let $X_1, ..., X_k$ be linearly independent vector fields at each point and verifying :

$$\begin{cases} \alpha(X_i) = 0, \ i = 1, \ldots, k, \\ [X_i, X_j] = \sum_{t=1}^{k} C_{ij}^t X_t. \end{cases}$$

According to Frobenius' theorem the distribution defined by these vector fields is completely integrable. It is contained in the kernel of the system. But if $s \geq k$ then

$$X_1 \rfloor X_2 \rfloor ... \rfloor X_k \rfloor \alpha \wedge (d\alpha)^s =$$

$$s(s-1) ... (s-k)\, \alpha \wedge (X_1 \rfloor d\alpha) \wedge (X_2 \rfloor d\alpha) \wedge ... \wedge (X_k \rfloor d\alpha) \wedge d\alpha^{s-k}.$$

This expression is zero if $k > s$. Since $\alpha \wedge (d\alpha)^s$ is of maximum rank and the vector fields $X_1, ..., X_k$ are linearly independent, the exterior form

$$X_1 \rfloor X_2 \rfloor ... \rfloor X_k \rfloor \alpha \wedge (d\alpha)^s$$

is zero. The hypothesis $k > s$ cannot be satisfied.

Lemma 10 *Let (α) be a Pfaffian system of rank 1 and of maximum class $n = 2s + 1$ and whose kernel contains an integrable distribution of (maximum) dimension s. Then there exists a coordinates system*

$$(x_1, ..., x_s, y_1, ..., y_s, z)$$

such that:

$$\alpha = dz + x_1 dy_1 + ... + x_s dy_s.$$

Proof of lemma. Let $X_1, ..., X_s$ be independent vector fields generating the integrable distribution contained in $\ker(S)$. Considering a coordinate system $(u_1, ..., u_n)$ on $\mathbb{R}^n$ such that the vector fields $X_1, ..., X_s$ verify

$$X_i = \frac{\partial}{\partial u_i}$$

for all $i = 1, ..., s$. Since $\alpha(X_i) = 0$ the Pfaffian form α can be written as:

$$\alpha = a_{s+1} du_{s+1} + ... + a_{2s+1} du_{2s+1}$$

with respect to this coordinates system.

The Pfaffian form α is non-zero thus there exists i $(i = s+1, ..., 2s+1)$ such that a_i is a non-zero function $(a_i \neq 0)$. We can assume that $a_{s+1} \neq 0$. Dividing α by a_{s+1} (obviously we do not change the Pfaffian system $\{\alpha = 0\}$), we obtain an expression for α of the form

$$\alpha = du_{s+1} + b_{s+2} du_{s+2} + ... + b_{2s+1} du_{2s+1}.$$

But (α) is of class $2s + 1$, hence $\alpha \wedge d\alpha^s \neq 0$, which is equivalent to

$$du_{s+1} \wedge db_{s+2} \wedge db_{s+2} \wedge ... \wedge db_{2s+1} \wedge db_{2s+1} \neq 0.$$

This proves that the functions $u_{s+1}, b_{s+2}, b_{s+2}, ..., b_{2s+1}, u_{2s+1}$ form a local coordinate system such that α takes the expected form.

Finally for the demonstration of Darboux' theorem in the case where α is of maximum class one needs the following lemma :

Lemma 11 *Let (α) be a Pfaffian system of rank 1 and of maximum class $n = 2s + 1$. There exists an integrable distribution of rank s contained in $\ker \alpha$.*

Consider a (local) basis of Pfaffian forms $\{\alpha = \alpha_1, \alpha_2, ..., \alpha_{2s+1}\}$ such that

$$d\alpha_1 = \alpha_2 \wedge \alpha_3 + ... + \alpha_{2s} \wedge \alpha_{2s+1}$$

and let $\{X_1, X_2, ..., X_{2s+1}\}$ be its dual basis. We have $X_2 \rfloor d\alpha_1 = \alpha_3$ and locally $\dim(\ker \alpha_1 \cap \ker \alpha_3) = 2s - 1$. Let Z be a vector field contained in $\ker \alpha_1 \cap \ker \alpha_3$. This vector field can be written

$$Z = \rho_2 X_2 + \rho_4 X_4 + \rho_5 X_5 + ... + \rho_{2s+1} X_{2s+1}.$$

The equation $[Z, X_2] = 0$ is equivalent to the differential system :

$$
\begin{pmatrix} \dot{\rho}_2 \\ \dot{\rho}_4 \\ \dot{\rho}_5 \\ \vdots \\ \dot{\rho}_{2s+1} \end{pmatrix} = \begin{pmatrix} 0 & C_{42}^2 & C_{52}^2 & \cdots \cdots & C_{2s+1,2}^2 \\ 0 & C_{42}^4 & C_{52}^4 & \cdots \cdots & C_{2s+1,2}^4 \\ \vdots & & & & \vdots \\ 0 & C_{42}^{2s+1} & C_{52}^{2s+1} & \cdots & C_{2s+1,2}^{2s+1} \end{pmatrix} \cdot \begin{pmatrix} \rho_2 \\ \rho_4 \\ \rho_5 \\ \vdots \\ \rho_{2s+1} \end{pmatrix},
$$

where

$$ \dot{\rho}_i = X_2(\rho_i) $$

and

$$ [X_i, X_j] = \sum C_{ij}^k X_k. $$

Notice that according to the expression of $d\alpha_1$ one has necessarily

$$ C_{ij}^1 = 0 $$

for $(i,j) \neq (2,3), \ldots, (2s, 2s+1)$. The space of solutions of the previous differential system is of dimension $2s-1$, thus we can choose $2s-1$ independent fundamental solutions. But,

$$ (\rho_2 = 1, \, \rho_4 = \rho_5 = \ldots = \rho_{2s+1} = 0) $$

is a basis of solutions. To each of these solutions these corresponds a vector field $Z = \sum \rho_i X_i$ satisfying $[X_2, Z] = 0$. Therefore we can construct independent vector fields $(X_2, Z_2, Z_3, \ldots, Z_{2s-1})$ corresponding to the chosen fundamental solutions.

Let us remark that the distribution spanned by these vector fields is of rank $2s - 1$ and that it is contained in the kernel of α_1. It is similarly so for the distribution spanned by the brackets $[Z_i, Z_j]$, but this distribution is certainly not integrable. Nevertheless, there exists an integrable sub-distribution of rank 2 which is spanned, for example, by X_2 and Z_2. To understand the process of construction of the maximal integrable distribution we are going to determine a third vector field Z independent with X_2 and Z_2 satisfying

$$ [X_2, Z] = [Z_2, Z] = 0 \text{ and } \alpha_1(Z) = 0. $$

The distribution defined by

$$ (\ker \, \alpha_1) \cap \ker(X_2 \rfloor d\alpha_1) \cap \ker(Z_2 \rfloor d\alpha_1) $$

is of constant rank $2s-2$ (on the open where it is defined). Let Z be a vector field belonging to this distribution and independent of X_2 and Z_2. This vector field exists if $s > 2$. Amongst the vector fields Z_i defined above there are $(2s - 4)$ independent vector fields, for example, $Z_4, ..., Z_{2s-1}$, satisfying the conditions imposed to the vector field Z. We take

$$Z = \mu_4 Z_4 + ... + \mu_{2s-1} Z_{2s-1},$$

where μ_i are smooth functions satisfying $X_2(\mu_i) = 0$. The equation $[Z_2, Z] = 0$ defines a linear differential system of order $2s - 4$, and for each chosen fundamental solution (amongst $2s - 4$ fundamental solutions) there corresponds a vector field $\widetilde{Z}$ such that

$$\left[X_2, \widetilde{Z}\right] = \left[Z_2, \widetilde{Z}\right] = \alpha_1 \left(\widetilde{Z}\right) = 0.$$

Thus we have constructed $(2s - 4)$ vector fields denoted by $\widetilde{Z}_4, ..., \widetilde{Z}_{2s-1}$, and the distribution spanned by the vector fields $X_2, Z_2, \widetilde{Z}_4$ is of rank 3 is integrable, and is contained in the kernel of α_1. It is sufficient now to iterate this process. Assume that we have m vector fields $\left(X_2, Z_2, \widetilde{Z}_4, \widetilde{Z}_5, ..., \widetilde{Z}_m\right)$ assumed to be linearly independent at each point of the open set of the definition, and satisfying

$$\left[X_2, \widetilde{Z}_i\right] = \left[Z_2, \widetilde{Z}_i\right] = [X_2, Z_2] = 0 \text{ and } \alpha_1 \left(\widetilde{Z}_i\right) = 0,$$

then we construct an $(m + 1)$-th vector field by taking

$$Z = \sum_{i=m+1}^{2s-1} \rho_i Y_i,$$

where Y_i are defined by the solutions of the differential system associated with the equation $\left[\widetilde{Z}_{m-1}, Z\right] = 0$. Thus we can notice that the distribution constructed is of rank s, thus it is maximal.

This completes the proof of the lemma, and also the Darboux theorem in the case where the system is of maximum class.

The case where class $(\alpha) < n$ can be considered the same to the previous case, taking a local basis $Y_1, ..., Y_{n-2s-1}$ of the characteristic distribution $C(\alpha)$, and a coordinates system $(u_1, ..., u_{n-2s-1}, x_1, ..., x_{2s+1})$ such that $Y_i = \partial/\partial u_i$. The system $\{\alpha\}$ induces on the space $(x_1, ..., x_{2s+1})$ a Pfaffian system of class $2s + 1$, and in this space the system $\{\alpha\}$ possesses the Darboux expression. This completes the proof of the Darboux theorem. ∎

4.5.3 Consequences

All Pfaffian system of rank 1 and of the class $2s + 1$ are locally isomorphic to the system

$$\{dx_1 + x_2dx_3 + ... + x_{2s}dx_{2s+1} = 0\}.$$

Thus, the classification, up to local isomorphism, of Pfaffian systems of rank 1 and of class $2s + 1$ is reduced to an unique model.

4.6 The Darboux theorem with parameters

This Darboux theorem 'with parameters' gives a generalization of the previous theorem in the following situation: Let $\mathfrak{F}$ be a given foliation and let α be a Pfaffian form such that on each leaf the system $\{\alpha = 0\}$ is of constant class $2s + 1$. The restriction of this system to each leaf is reduced to the Darboux form. We are going to show that there exists a transverse coordinates system $(u_1, ..., u_{n-p}, x_1, ..., x_p)$ such that the foliation is defined by the equation $u_i = $ constant. The induced form on each leaf can, by Darboux' theorem be expressed as a function of the coordinates $x_1, ..., x_{n-p}$.

Theorem 4.5 *Let $\mathfrak{F}$ be a local foliation on $\mathbb{R}^n$ of dimension p and let $\{\alpha\}$ be a Pfaffian system of rank 1 such that the restriction to each leaf defines a Pfaffian system of constant class $2s+1$. Then, there exists a local coordinates system $(u_1, ..., u_{n-p}, x_1, ..., x_p)$ such that each leaf is defined by the equations*

$$du_1 = 0, ..., du_{n-p} = 0$$

whose restriction of the Pfaffian form α to each leaf is written

$$\alpha = dx_1 + x_2dx_3 + ... + x_{2s}dx_{2s+1}.$$

Proof. Since $\mathfrak{F}$ is a p-dimensional foliation on $\mathbb{R}^n$ we can find a coordinate system $(u_1, ..., u_{n-p}, v_1, ..., v_p)$ such that the leaves of $\mathfrak{F}$ are parametrized by the equations

$$\begin{cases} u_1 = \text{ constant,} \\ \quad \vdots \\ u_{n-p} = \text{ constant.} \end{cases}$$

Locally, each leaf is a $p-$plane. With respect to this coordinates system the Pfaffian form α can be written as:

$$\alpha = \sum_{i=1}^{p} a_idv_i + \tilde{\alpha},$$

where a_i are smooth functions of the variables $u_1, ..., u_{n-p}, v_1, ..., v_p$, and $\tilde{\alpha} = \sum b_i du_i$. The Pfaffian form $\tilde{\alpha}$ is zero in restriction to each leaf. Hence, $\alpha_1 = \sum a_i dv_i$ induces a Pfaffian form of class $2s+1$ on each leaf. We consider now the differential form

$$\alpha_1 = \sum a_i (u_1, ..., u_{n-p}, v_1, ..., v_p) \, dv_i$$

as a Pfaffian form on the space defined by the coordinates $(v_1, ..., v_p)$, the functions u_i playing the role of parameters. For that reason this theorem will be termed 'Darboux' theorem with parameters'). We resume now the proof of Darboux' theorem, point by point, taking into account these new parameters. We always assume first, that each leaf is of dimension $2s + 1$ and class $(\alpha_1) = 2s + 1$.

Lemma 12 *Let D be an integral distribution of α_1 and tangent to each leaf. Then* $\text{rank}\,(D) \le s$.

In fact, by restriction to each leaf we return to the situation of the previous paragraph.

Lemma 13 *If there exists an integral distribution of rank s and tangent to each leaf, then we can find a local coordinates system*

$$(x_1, ..., x_{2s+1}, u_1, ..., u_{n-2s-1})$$

such that

$$\alpha_1 = dx_1 + x_2 dx_3 + ... + x_{2s} dx_{2s+1}.$$

Proof of lemma. Let $(X_1, ..., X_s)$ be a tangent distribution to each leaf and satisfying $\alpha_1 (X_i) = 0$. We can find coordinates

$$(v_1, ..., v_{2s+1}, u_1, ..., u_{n-2s-1})$$

such that the leaves of the initial foliation are defined by equations $u_i = $ constant, the vector fields X_i are given by $X_i = \partial/\partial x_i$ $(i = 1, ..., s)$. With respect to this coordinates system, we have

$$\alpha_1 = a_{s+1} dx_{s+1} + ... + a_{2s+1} dx_{2s+1}.$$

We can assume that $a_{s+1} = 1$, that is $\alpha_1 = dx_{s+1} + ... + a_{2s+1} dx_{2s+1}$. The condition $\alpha_1 \wedge (d\alpha_1)^s \neq 0$ shows that the functions

$$(x_{s+1}, ..., x_{2s+1}, a_{s+2}, ..., a_{2s+1})$$

are independent. This proves the lemma.

The lemma 11 of the previous section is unchanged. Indeed, the key to the proof is based on the existence of independent solutions of a linear differential system. The existence of parameters does not change anything in this proof.

Finally, we examine the case where α_1 is not of maximum class. Then, we foliate each leaf of the foliation $u_i = $ constant by the characteristic foliation of α_1, which is tangent to $(u_i = $ constant) by the construction of α_1.

This completes the proof of the Darboux' theorem with parameters. ∎

4.7 Invariants of Pfaffian systems

The classification up to local isomorphism of Pfaffian systems of constant rank is based essentially on the study of invariants of a Pfaffian system. We have defined above a fundamental invariant, *the class*. In the case of Pfaffian systems of rank 1 this invariant entirely determines the (local) classification by the Darboux theorem. Similarly, the Pfaffian systems of constant rank r and constant class r are locally isomorphic; that is a consequence of Frobenius' theorem. In the other cases the classification is less easy to establish. Here, we are going to study some classic invariants so as to be able to undertake classifications in low dimensions.

4.7.1 Derived system of a Pfaffian system

Let (S) be a Pfaffian system of rank r in $\mathbb{R}^n$.

Definition 4.10 *The first derived system $D^1(S)$ of (S) is the Pfaffian system defined by*

$$D^1(S) = \{\alpha \in (S) \setminus d\alpha = 0 \mod (S)\}.$$

Recall that $d\alpha = 0 \mod(S)$ means that $d\alpha \wedge \alpha_1 \wedge ... \wedge \alpha_r = 0$, where $(\alpha_1, ..., \alpha_r)$ is a local basis of (S). By definition $D^1(S)$ is a Pfaffian subsystem of (S), that is, for each point x $\ker_x(D^1(S))$ contains $\ker_x(S)$. Thus we write $D^1(S) \subset (S)$.

Let us consider the following examples:

1. Let (S) be a Pfaffian system in $\mathbb{R}^4$ defined by

$$\begin{cases} \alpha_1 = dy_3 + y_1 dy_4, \\ \alpha_2 = dy_2 + y_3 dy_4. \end{cases}$$

Then $\{\alpha_2\}$ is a basis of $D^1(S)$. In fact, we have

$$d\alpha_2 = dy_3 \wedge dy_4,$$
$$\alpha_1 \wedge \alpha_2 = dy_3 \wedge dy_2 + y_3 dy_3 dy_4 + y_1 dy_4 dy_2,$$
$$d\alpha_2 \wedge \alpha_1 \wedge \alpha_2 = 0.$$

2. Let (S) be the Pfaffian system in $\mathbb{R}^5$ defined by

$$\begin{cases} \alpha_1 = dy_1 + y_3 dy_4, \\ \alpha_2 = dy_2 + y_3 dy_5. \end{cases}$$

Here we have $D^1(S) = \{0\}$.

Remark 7

The derived system $D^1(S)$ is not of constant rank in the general case. In the approach to the classification of Pfaffian systems we will assume that $D^1(S)$ is of constant rank. We can always return to this case by considering a neighborhood where $D^1(S)$ is of maximal rank. When $D^1(S)$ is not trivial it is not completely integrable in the general case. The definition of this system makes obvious a notion of integrability modulo (S) and not modulo $D^1(S)$. By the definition of $D^1(S)$ this system is an invariant: two locally isomorphic Pfaffian systems (S) and (S') have locally isomorphic derived systems $D^1(S)$ and $D^1(S')$. Thus the rank of $D^1(S)$ and the class of $D^1(S)$ are numerical invariants of (S).

4.7.2   The derived systems role

Let (S) be a Pfaffian system of constant rank r and of constant class s. Assume that the derived system $D^1(S)$ is of constant rank. Then we have:

Proposition 4.4 *Under the previous hypothesis the characteristic space of $D^1(S)$ contains the characteristic space of (S) at each point.*

Proof. In fact, considering independent Pfaffian forms

$$\alpha_1, \cdots, \alpha_d, \beta_{d+1}, \cdots, \beta_r$$

such that

$$(S) = \left(\alpha_1, \cdots, \alpha_d, \beta_{d+1}, \cdots, \beta_r\right)$$

and

$$D^1(S) = (\alpha_1, \cdots, \alpha_d).$$

These forms satisfy

$$d\alpha_i = \sum_{j=1}^{d} \alpha_j \wedge \gamma_j^i + \sum_{j=d+1}^{r} \beta_j \wedge \zeta_j^i \qquad (i = 1, \ldots, d),$$

$$d\beta_i = \sum_{j=1}^{d} \alpha_j \wedge \eta_j^i + \sum_{j=d+1}^{r} \beta_j \wedge \xi_j^i + \theta_i, \qquad (i = 1, \ldots, d),$$

where θ_i is a differential 2-form which does not belong to the ideal spanned by (S). Let $Y \in C(S)$. This vector satisfies

$$\begin{cases} \alpha_i(Y) = \beta_j(Y) = 0, \\ Y \rfloor d\alpha_i = Y \rfloor d\beta_j = 0 \quad \mathrm{mod}(S). \end{cases}$$

We prove that $Y \rfloor d\alpha_i = 0 \ \mathrm{mod}(D(S))$. We know that $d(d\alpha_i) = 0$. Differentiating the expression of $d\alpha_i$, we obtain:

$$0 = \sum_{j=d+1}^{r} d\beta_j \wedge \zeta_j^i \quad \mathrm{mod}(S),$$

then

$$\begin{aligned}
0 &= Y \rfloor \left(\sum_{j=d+1}^{r} d\beta_j \wedge \zeta_j^i \right) \quad \mathrm{mod}(S) \\
&= \sum_{j=d+1}^{r} Y \rfloor d\beta_j \wedge \zeta_j^i + d\beta_j \wedge Y \rfloor \zeta_j^i \quad \mathrm{mod}(S) \\
&= \sum_{j=d+1}^{r} d\beta_j \wedge Y \rfloor \zeta_j^i \quad \mathrm{mod}(S) \\
&= \sum_{j=d+1}^{r} \theta_j \wedge Y \rfloor \zeta_j^i \quad \mathrm{mod}(S).
\end{aligned}$$

Thus

$$\begin{aligned}
Y \rfloor d\alpha_i &= \sum_{j=d+1}^{r} \beta_j \wedge \zeta_j^i(Y) \quad \mathrm{mod}(D(S)) \\
Y \rfloor d\beta_i &= Y \rfloor \theta_i = 0 \quad \mathrm{mod}(S).
\end{aligned}$$

Since θ_i does not belong to the ideal spanned by (S), we necessarily have $Y \rfloor \theta_i = 0$. We deduce that $Y \rfloor \zeta_j^i = 0$; otherwise one of the Pfaffian forms β_i would be in $D(S)$. Thus

$$Y \rfloor d\alpha_i = \sum_{j=d+1}^{r} \beta_j \wedge \zeta_j^i(Y) = 0 \quad \mathrm{mod}(D(S)).$$

This proves the proposition. ∎

This means that the number of differentials or variables necessary for writing the derived system is less than that of the initial Pfaffian system. The operation of taking the derived system is, in a way, an operation of eliminating variables. Notice, finally, that the characteristic system of (S) (that is the Pfaffian system having for kernel the characteristic spaces of (S)) is, in a way, the smallest completely integrable Pfaffian system containing (S). The derived system plays an opposed role. One seeks inside (S) a completely integrable sub-system. It is for example, the last derived system not reduced to $\{0\}$ when (S) is totally regular. We will define the last notion in the next paragraphs.

4.7.3　Successive derived systems

Definition 4.11 *Let (S) be a Pfaffian system of constant rank and let $D^1(S)$ be its first derived system. If $D^1(S)$ is of constant rank, the first derived system $D^1(D^1(S))$ of $D^1(S)$ will be called the second derived system of (S) and denoted by $D^2(S)$.*

In the general case, if the kth derived system $D^k(S)$ is of constant rank, we define the $(k+1)^{th}$ derived system by:

$$D^{k+1}(S) = D^1\left(D^k(S)\right).$$

If the kth derived system is well defined we have the decreasing sequence

$$D^k(S) \subset D^{k-1}(S) \subset \ldots \subset D^1(S) \subset (S).$$

Proposition 4.5 *If $D^k(S) = D^{k-1}(S)$ then the Pfaffian system $D^{k-1}(S)$ is completely integrable and $D^{k+p}(S) = D^{k-1}(S)$ for every $p \geq 0$.*

Proof. In fact, $D^1(S) = (S)$ means that

$$(S) = \{\alpha \in S \setminus d\alpha = 0 \ \mathrm{mod}(S)\}.$$

According to Frobenius' theorem (S) is completely integrable. ∎

4.7.4   Totally regular systems

Definition 4.12 *We say that* (S) *is a totally regular system if all successive derived systems are of constant rank.*

If (S) is totally regular we can consider the strictly decreasing sequence of Pfaffian systems

$$(S) \supset D^1(S) \supset D^2(S) \supset \ldots \supset D^k(S),$$

the integer k is defined by the condition that $D^k(S) = D^{k+p}(S) \neq \{0\}$ for all p or $D^k(S) = \{0\}$ and $D^{k-1}(S) \neq \{0\}$. This integer k is called the *length* of (S). Two locally isomorphic totally regular systems are the same length. We say that a totally regular system (S) with length k is of the *first type* if $D^k(S) = \{0\}$, and of the *second type* if $D^k(S) = D^{k+p}(S) \neq \{0\}$, that is, if $D^k(S)$ is of non-zero constant rank and is completely integrable.

Remark 8 *Suppose that* (S) *is of the second type and of length* k, *and take* $t = \mathrm{rank}\,(D^k(S))$. *On each integral manifold* F_{n-t} *of* $D^k(S)$ *the system* (S) *induces a totally regular system of the first type.*

An interesting class of totally regular systems is the class of flag systems defined as follows:

Definition 4.13 *A totally regular Pfaffian system* (S) *with length* k *is called a flag system if the following condition is satisfied:*

$$\mathrm{rank}\,(D^p(S)) = \mathrm{rank}\,(D^{p+1}(S)) + 1,$$

for each $p \leq k$.

Example 14

Consider the Pfaffian system (S) in $\mathbb{R}^5$ defined by the following forms:

$$\begin{cases} \alpha_1 = dy_1 + y_2 dy_5, \\ \alpha_2 = dy_2 + y_3 dy_5, \\ \alpha_3 = dy_3 + y_4 dy_5. \end{cases}$$

This system is of rank 3 and of class 5, and it satisfies :

$$\begin{cases} d\alpha_1 = dy_2 \wedge dy_5, \\ d\alpha_2 = dy_3 \wedge dy_5, \\ d\alpha_3 = dy_4 \wedge dy_5. \end{cases}$$

Hence

$$\begin{aligned} \alpha_1 \wedge \alpha_2 \wedge \alpha_3 \ = \ & dy_1 \wedge dy_2 \wedge dy_3 + y_4 dy_1 \wedge dy_2 \wedge dy_5 + y_3 dy_1 \wedge dy_5 \wedge dy_3 \\ & + y_2 dy_5 \wedge dy_2 \wedge dy_3. \end{aligned}$$

Let $\beta = a_1 \alpha_1 + a_2 \alpha_2 + a_3 \alpha_3$. We have

$$d\beta = da_1 \wedge \alpha_1 + da_2 \wedge \alpha_2 + da_3 \wedge \alpha_3 + a_1 d\alpha_1 + a_2 d\alpha_2 + a_3 d\alpha_3$$

and

$$d\beta = 0 \mod (S) \iff a_1 d\alpha_1 + a_2 d\alpha_2 + a_3 d\alpha_3 = 0 \mod (S).$$

This is equivalent to $a_3 = 0$. Thus $D^1(S) = \{\alpha_1, \alpha_2\}$ and this derived system is of rank 2. Similarly,

$$D^2(S) = \{\beta \in D^1(S) \setminus d\beta = 0 \mod(D^1(S))\} = \{\alpha_1\}$$

and rank $D^2(S) = 1$. Finally, $D^3(S) = \{0\}$. Thus (S) is a flag system.

Proposition 4.6 *Let (S) be a flag system of rank r and of length k, $k \geq 2$. Then the successive derived systems $D^1(S), ..., D^k(S)$ are of constant classes, and we have*

$$\begin{cases} \text{class } \left(D^k(S) \right) = \text{ rank } \left(D^k(S) \right), \\ \text{class} \left(D^p(S) \right) = \text{ rank } \left(D^p(S) \right) + 2 \, , \, 0 < p < k. \end{cases}$$

Proof. Since $D^k(S)$ is zero or completely integrable, we certainly have

$$\text{class} \left(D^k(S) \right) = \text{rank}(D^k(S)).$$

Let $\{\alpha_1, ..., \alpha_r\}$ be a local basis of (S) such that $\{\alpha_1, ..., \alpha_p\}$ is a basis of $D^{r-p}(S)$. Since (S) is a flag system, we have together $k = r$ and

$$\begin{cases} d\alpha_1 = 0 \mod (S), \\ \quad \vdots \\ d\alpha_{r-1} = 0 \mod (S), \\ d\alpha_r \wedge \alpha_1 \wedge \cdots \wedge \alpha_r \neq 0. \end{cases}$$

Then

$$da_i = \sum_{j=1}^{r} \alpha_j \wedge \beta_j^i, \quad i = 1, \ldots, r-1,$$

$$da_r = \sum_{j=1}^{r} \alpha_j \wedge \beta_j^r + \theta \quad \text{with } \theta \neq 0 \mod (S).$$

We deduce that class $(S) = r+ \text{rank } (\theta)$ (which justifies the hypothesis $k \geq 2$). Since $\{\alpha_1, \ldots, \alpha_{r-1}\}$ is a local basis of $D^1(S)$, $(\alpha_1, \cdots, \alpha_{r-2})$ being a local basis of $D^2(S)$, we have

$$\begin{cases} d\alpha_1 = 0 \mod (D^1(S)), \\ \quad \vdots \\ d\alpha_{r-2} = 0 \mod (D^1(S)), \\ d\alpha_{r-1} \wedge \alpha_1 \wedge \cdots \wedge \alpha_{r-1} \neq 0. \end{cases}$$

But $d\alpha_{r-1} \wedge \alpha_1 \wedge \cdots \wedge \alpha_r = 0$ which implies

$$d\alpha_{r-1} = \sum_{j=1}^{r-1} \alpha_j \wedge \beta_j^{r-1} + \alpha_r \wedge \beta_r^{r-1}$$

with $\beta_r^{r-1} \neq 0$. Hence

$$\text{class} \left(D^1(S) \right) = r - 1 + 2 = r + 1 = \text{rank} \left(D^1(S) \right) + 2.$$

In general, let $\{\alpha_1, \cdots, \alpha_{r-k}\}$ be the basis of $D^k(S)$ such that the forms $\{\alpha_1, \cdots, \alpha_{r-k-1}\}$ define a basis of $D^{k+1}(S)$. We have

$$\begin{cases} d\alpha_1 = 0 \mod (D^k(S)), \\ \quad \vdots \\ d\alpha_{r-k-1} = 0 \mod (D^k(S)), \\ d\alpha_{r-k} \wedge \alpha_1 \wedge \cdots \wedge \alpha_{r-k} \neq 0, \\ d\alpha_{r-k} \wedge \alpha_1 \wedge \cdots \wedge \alpha_{r-k} \wedge \alpha_{r-k+1} = 0. \end{cases}$$

We deduce that

$$d\alpha_{r-k} = \sum_{j=1}^{r-k} \alpha_j \wedge \beta_j^{r-k} + \alpha_{r-k+1} \wedge \beta_{r-k+1}^{r-1}$$

with $\beta_{r-k+1}^{r-k} \neq 0$ thus

$$\text{class} \left(D^k(S) \right) = r - k + 2.$$

This completes the proof of the proposition. ∎

4.7.5   The Engel invariant. The Gardner inequalities

Let (S) be a Pfaffian system of rank r and of constant class c.

Definition 4.14 *The Engel invariant $e\,(S)$ of (S) is the smallest integer n such that $(d\omega)^{n+1} = 0 \bmod (S)$ for each $\omega \in (S)$.*

Two locally isomorphic Pfaffian systems have the same Engel invariant. When $e\,(S) = 0$ the Pfaffian system (S) is completely integrable.

Example 15

Considering the following examples:

1. Let (α_1, α_2) be a Pfaffian system in $\mathbb{R}^5$ defined by

$$\begin{cases} \alpha_1 = dx_1 + x_3 dx_4, \\ \alpha_2 = dx_2 + x_3 dx_5. \end{cases}$$

Then $e\,(S) = 1$. In fact, we have $\alpha_1 \wedge \alpha_2 \neq 0$ and $(d\alpha_i)^2 \wedge \alpha_1 \wedge \alpha_2 = 0$ for every $i = 1, 2$.

2. Let $(\alpha_1, \alpha_2, \alpha_3)$ be a Pfaffian system in $\mathbb{R}^7$ defined by

$$\begin{cases} \alpha_1 = dx_1 + x_4 dx_5 + x_6 dx_7, \\ \alpha_2 = dx_2 + x_4 dx_6 - x_5 dx_7, \\ \alpha_3 = dx_3 + x_4 dx_7 + x_5 dx_6. \end{cases}$$

In this case we have $e\,(S) = 3$.

Proposition 4.7 *Let (S) be a Pfaffian system of rank r and of constant class c of which the Engel invariant is $e\,(S)$. Then we have*

$$\frac{c - r}{r + 1} \leq e\,(S) \leq \frac{c - r}{2},$$

called the Gardner inequalities.

Proof. Let $(\alpha_1, \ldots, \alpha_r)$ be a local basis of (S), which we extend to a Pfaffian system $(\alpha_1, \cdots, \alpha_r, \beta_1, \cdots, \beta_{n-r})$ of maximum rank n. The differentials $d\alpha_i$ can be written

$$d\alpha_i = \sum a^i_{jk} \alpha_j \wedge \alpha_k + \sum b^i_{jk} \alpha_j \wedge \beta_k + \sum c^i_{jk} \beta_j \wedge \beta_k.$$

Since the Pfaffian system (S) is of class c, the rank of the Pfaffian system

$$\left(\sum_{j=1}^{n-r} c^i_{jk} \beta_j, \quad i = 1, \cdots, n - r \right)$$

is less than $c - r$. It follows that for all i the 2-form

$$\sum c^i_{jk} \beta_j \wedge \beta_k$$

is of rank less or equal to $(c-r)/2$. Then $(d\alpha_i)^{[(\frac{c-r}{2})]+1} = 0 \bmod (S)$ for each i. There $[a]$ denotes the integer part of a. The choice of the basis $(\alpha_1, \cdots, \alpha_r)$ being arbitrary, one has $e\,(S) \le (c-r)/2$. The second inequality results from the following remark:if $(X_1, \cdots, X_r, Y_1, \cdots, Y_{n-r})$ is a distribution of rank n verifying

$$\beta_i\,(Y_j) = \delta_{ij}$$

and

$$\alpha_i\,(Y_j) = \beta_i\,(X_j) = 0,$$

then the Pfaffian system

$$\left(Y_j \rfloor d\alpha_i = \sum c^i_{jk} \beta_k - \sum b^i_{kj} \alpha_k \right)$$

is of rank c. Consequently the Pfaffian system

$$\left(\sum_{j=1}^{n-r} c^i_{jk} \beta_k \; (i = 1, \cdots, r \; ; \; j = 1, \cdots, n - r) \right)$$

is of rank $c - r$ and there exists $\alpha = \sum a_i \alpha_i$ such that

$$(d\alpha)^{\frac{(c-r)}{(r+1)}} \ne 0 \bmod (S).$$

Then, $e\,(S) \ge (c - r)/(r + 1)$ it is the second Gardner inequality. ∎

4.8 The Schouten - Van Der Kulk theorem

Theorem 4.6 *Let (S) be a Pfaffian system of constant rank r and of constant class c, whose Engel invariant $e\,(S)$ is m. There exists a local basis $(\alpha_1, \cdots, \alpha_r)$ of (S) such that each Pfaffian form α_i is of class $2m + 1$.*

Let us note that in general it is not easy to find the proposed basis in the Shouten-Van Der Kulk' theorem. We dot not propose the proof of this theorem. We can refer to the one given by E. Cartan which is based on the notion of involutive Pfaffian system.

Example 16

Let us consider the following examples:

1. We take $n = 7$; $r = 3$ and $c = 7$. Considering the system (S) given by:
$$\begin{cases} \alpha_1 = dx_1 + x_4 dx_5 + x_6 dx_7, \\ \alpha_2 = dx_2 + x_4 dx_6 - x_5 dx_7, \\ \alpha_3 = dx_3 + x_4 dx_7 + x_5 dx_6. \end{cases}$$
Here we have $e(S) = 2$. According to the Gardner inequalities, we have $e(S) \leq (7-3)/2 = 2$. Then we have an example of a Pfaffian system of which Engel invariant is maximal. The Pfaffian forms α_1, α_2, and α_3 are of class $5 = 2 \times 2 + 1$.

2. Consider now the following Pfaffian system given by
$$\begin{cases} \alpha_1 = dx_1 + x_4 dx_5, \\ \alpha_2 = dx_2 + x_4 dx_6, \\ \alpha_3 = dx_3 + x_4 dx_7. \end{cases}$$

As previously, we have $n = c = 7$ and $r = 3$. But here $e(S) = 1$. Thanks to the Gardner inequalities we find that the possible minimum for such systems is $(c-2)/(r+1) = 1$. This is an example in which the Engel invariant is minimal. We also note that the basis forms satisfy the Schouten and Van der Kulk theorem.

4.9   Maximal integral manifolds

Consider a Pfaffian system (S) of constant rank r and of constant class c. If this system is not completely integrable it is reasonable to be interested in the nature of integral manifolds of the Pfaffian system, and especially in integral manifolds of maximum dimension. We have seen, in the case where (S) is of rank 1 and of class $c = 2s + 1$, that there exists a tangent foliation of the kernel of (S) and of maximal dimension $s + (r - c)$. The case where the rank is strictly greater than 1, is certainly not regular: the datum of a constant rank and a constant class does not allow us to ensure the existence of an integral foliation of (S) is of maximal dimension. We propose in this paragraph to study the upper boundary of the dimension of tangent manifolds of $\ker(S)$ in order to see how the Darboux foliations must be in the case where $r > 1$.

Theorem 4.7 *Let (S) be a Pfaffian system of rank r and of maximum class n. Then every local tangent foliation to the distribution $\ker(S)$ is of dimension d which satisfies*

$$d \le \frac{r(n-r)}{r+1}.$$

Proof. Let $(\alpha_1, \cdots, \alpha_r)$ be a local basis of (S) and consider a distribution of rank n spanned by $(X_1, \cdots, X_r, Y_1, \cdots, Y_q, Z_1, \cdots, Z_s)$, where $r + q + s = n$, satisfying

$$\begin{cases} \alpha_i(X_j) = \delta_{ij}, \\ \alpha_i(Y_j) = \alpha_i(Z_j) = 0 \end{cases}$$

at each point.

Assume that the distribution generated by $(Y_1, \cdots, Y_q)$ is integrable, that is, $[Y_i, Y_j] = \sum C_{ij}^k Y_k$. We denote by $(\alpha_1, \cdots, \alpha_r, \beta_1, \cdots, \beta_q, \gamma_1, \cdots \gamma_s)$ the dual forms to (X_i, Y_j, Z_k). We can write:

$$\begin{aligned} d\alpha_i &= \sum a_{jk}^i \alpha_j \wedge \alpha_k + \sum b_{jk}^i \alpha_j \wedge \beta_k + \sum c_{jk}^i \alpha_j \wedge \gamma_k \\ &+ \sum d_{jk}^i \beta_j \wedge \beta_k + \sum e_{jk}^i \beta_j \wedge \gamma_k + \sum f_{jk}^i \gamma_j \wedge \gamma_k. \end{aligned}$$

Since the distribution spanned by $(Y_1, \cdots, Y_q)$ is integrable, the functions d_{jk}^i are zero identically. The system (S) being of maximum class, the Pfaffian system spanned by the forms

$$Z_j \rfloor d\alpha_i = \sum c_{jk}^i \alpha_k + \sum e_{jk}^i \beta_k + \sum f_{jk}^i \gamma_k.$$

is of rank $\ge q$. But it is a Pfaffian system of order sr, therefore we have the inequality $sr \ge q$. Hence $(n - r - q)r \ge q$, this is equivalent to $q \le r(n-r)/(r+1)$. This proves the theorem. ∎

Remark 9

We observe that when $r = 1$ we have $q \leq (n-1)/2$, that is $q \leq s$ (because in that case $n = 2s + 1$). But, contrarily to this situation, where we have made obvious the existence of a foliation of dimension $(n-1)/2$, one cannot assert anything, in the general case, about the existence or not of a foliation of dimension $r(n-r)/(r+1)$. For example, the Pfaffian system in $\mathbb{R}^7$

$$
\begin{cases}
\alpha_1 = dx_1 + x_4 dx_5, \\
\alpha_2 = dx_2 + x_4 dx_6, \\
\alpha_3 = dx_3 + x_4 dx_7
\end{cases}
$$

admits an integral foliation of dimension 3. This foliation is spanned by the vector fields

$$
Y_1 = \frac{\partial}{\partial x_5} - x_4 \frac{\partial}{\partial x_1}, \quad Y_2 = \frac{\partial}{\partial x_6} - x_4 \frac{\partial}{\partial x_2} \quad \text{and} \quad Y_3 = \frac{\partial}{\partial x_7} - x_4 \frac{\partial}{\partial x_3}
$$

and we have $q = 3 = 3.(7-3)/(3+1)$.

In this example, the Pfaffian system is of maximum class and its kernel is tangent to a three-dimensional foliation. On the other hand, the Pfaffian system given by

$$
\begin{cases}
\alpha_1 = dx_1 + x_4 dx_5 + x_6 dx_7, \\
\alpha_2 = dx_2 + x_4 dx_6 - x_5 dx_7, \\
\alpha_3 = dx_3 + x_4 dx_7 + x_5 dx_6,
\end{cases}
$$

although being of rank 3 and of maximal class 7 like the previous one, it is not endowed with an integral foliation of dimension 3. One can see that, in fact, the tangent foliations of the kernel of (S) are at the most of dimension 1.

Proposition 4.8 *Let (S) be a Pfaffian system of rank r and of maximum class $n = c$ endowed with an integral foliation of maximal dimension $q = r(n-r)/(r+1)$. Then*

1. *$n = p + r + rp$;*

2. *The Engel invariant of (S) is equal to $(n-r)/(r+1)$.*

Proof. In a straightforward way we take $q = pr$; we have $n - r = q + q/r$. The Engel invariant can be calculated easily enough. It will be sufficient to notice that for every form α of the Pfaffian system (S) we have $d\alpha(Y_i, Y_j) = 0$, where Y_i and Y_j are tangent vector fields of the foliation. But the dimension of the foliation is equal to $r(n-r)/(r+1)$ from which one deduces that $(d\alpha)^{p+1} = 0 \bmod (S)$. ■

The inequality given in the preceding theorem can be generalized to the case of those systems of constant class c: let (S) be a Pfaffian system of constant class c and of maximum rank r. Then every integral local foliation of (S) is at most of dimension $q = r(c-r)/r+1)$.

The proof is identical to the preceding, the single change resides in the consideration of the characteristic foliation, which is locally transverse to the integral foliation.

Chapter 5

CLASSIFICATION OF PFAFFIAN SYSTEMS

We are going, in this chapter, to give all the local models, up to isomorphism, of Pfaffian systems of constant rank and class on $\mathbb{R}^n$ with $3 \leq n \leq 5$. Let us point out some conventions and notations used in the preceding chapters. Let (S) be a Pfaffian system defined by the equations

$$\left\{ \begin{array}{c} \omega_1 = 0, \\ \vdots \\ \omega_r = 0. \end{array} \right.$$

The Pfaffian forms ω_i are assumed to be independent at any point and, to simplify the writing of this system, one writes

$$(S) = \{\omega_1, \cdots, \omega_r\}$$

or

$$(S) = \left\{ \begin{array}{c} \omega_1, \\ \vdots \\ \omega_r \end{array} \right. .$$

One also says that these forms constitute a basis of (S), and if α is a linear combination of the basis forms ω_i, one writes $\alpha \in (S)$.

5.1 Pfaffian systems on $\mathbb{R}^3$

Let (S) be a Pfaffian system with three variables.

1. If the rank of (S) is one, the class c of this system is either 1 or 3. Thus one has, according to the theorems of Darboux and Frobenius, the two following models:

$$(S_3^1) = \{\omega_1 = dy_1 \qquad\qquad r = 1, c = 1$$

$$(S_3^2) = \{\omega_1 = dy_1 - y_2 dy_3 \quad r = 1, c = 3.$$

Every Pfaffian system of rank 1 and constant class is locally isomorphic to (S_3^1) or (S_3^2).

2. The rank of (S) is 2. The class of (S) is either 2 or 3. As there is no Pfaffian system on $\mathbb{R}^n$ of rank $n-1$ and from class n, one deduces from it that any Pfaffian system on $\mathbb{R}^3$ of rank 2 is of class 2. Such a system is completely integrable and locally isomorphic to

$$(S_3^3) = \begin{cases} \omega_1 = dy_1, \\ \omega_2 = dy_2. \end{cases}$$

3. The rank of (S) is 3. It is thus completely integrable and by the Frobenius' theorem it is locally isomorphic to

$$(S_3^4) = \begin{cases} \omega_1 = dy_1, \\ \omega_2 = dy_2, \\ \omega_3 = dy_3. \end{cases}$$

Theorem 5.1 *Every Pfaffian System on $\mathbb{R}^3$ of constant rank and class is locally isomorphic to one of the following system described in Table 1. These systems are pairwise not isomorphic.*

TABLE 1

Pfaffian systems on $\mathbb{R}^3$

$(S_3^1) = \{\omega_1 = dy_1$	$r = 1,$	$c = 1$
$(S_3^2) = \{\omega_1 = dy_1 - y_2 dy_3$	$r = 1,$	$c = 3$
$(S_3^3) = \begin{cases} \omega_1 = dy_1 \\ \omega_2 = dy_2 \end{cases}$	$r = 2,$	$c = 2$
$(S_3^4) = \begin{cases} \omega_1 = dy_1 \\ \omega_2 = dy_2 \\ \omega_3 = dy_3 \end{cases}$	$r = 3,$	$c = 3$

5.2 Pfaffian Systems on $\mathbb{R}^4$

5.2.1 Pfaffian systems of rank 1, 3, 4

Case $r = 4$

If $r = 4$ the system (S) is completely integrable and, by Frobenius' theorem, it admits the following local model:

$$(S_4^0) = \begin{cases} \omega_1 = dx_1, \\ \omega_2 = dx_2, \\ \omega_3 = dx_3, \\ \omega_4 = dx_4. \end{cases}$$

Case $r = 3$

The system (S) is completely integrable and locally isomorphic to the following system:

$$(S_4^1) = \begin{cases} \omega_1 = dx_1, \\ \omega_2 = dx_2, \\ \omega_3 = dx_3. \end{cases}$$

Case $r = 1$

One is brought back to the Darboux theorem. The system (S) is then locally isomorphic to the one of the following:

$$(S_4^2) = \{\omega_1 = dx_1 + x_2 dx_3$$

if $c = 3$ and

$$(S_4^3) = \{\omega_1 = dx_1$$

if $c = 1$.

5.2.2 Pfaffian systems of rank 2

class$(S) = 2$

Since the rank is equal to the class, such a system is completely integrable. It is written locally

$$(S_4^4) = \begin{cases} \omega_1 = dy_1 \\ \omega_2 = dy_2. \end{cases}$$

class$(S) \geq 3$

Let (ω_1, ω_2) be a basis of (S). These forms satisfy

$$\begin{cases} d\omega_1 = A\omega_3 \wedge \omega_4 \\ d\omega_2 = B\omega_3 \wedge \omega_4 \end{cases} \mod (S) ,$$

the Pfaffian forms ω_1, ω_2, ω_3, ω_4 being linearly independent at any point. As the Pfaffian system (S) is not completely integrable one of the two functions A or B is not identically zero. Let us suppose that $A(x) \neq 0$ at any point x (the common singularities of A and B are not considered here for they imply singularities on the class of (S)). Let us consider a change of basis of (S):

$$\begin{cases} \bar{\omega}_1 = \frac{1}{A}\omega_1, \\ \bar{\omega}_2 = \omega_2 - \frac{B}{A}\omega_1. \end{cases}$$

The Pfaffian equations $\bar{\omega}_1 = \bar{\omega}_2 = 0$ define the system (S) and satisfy

$$\begin{cases} d\bar{\omega}_1 = \omega_3 \wedge \omega_4 \\ d\bar{\omega}_2 = 0 \end{cases} \mod (S) .$$

The system (S) is thus a Pfaffian system on $\mathbb{R}^4$ of rank 2 and constant class equal to 4. The first derived system of (S) is the Pfaffian system

$$D^1 (S) = \{\bar{\omega}_2\}.$$

It is of rank 1. As for the class, two cases can arise:

- class $(D^1 (S)) = 1$ that is, $D^1 (S)$ is completely integrable,
- class $(D^1 (S)) = 3$.

(Here, again, we consider only Pfaffian systems whose numerical invariants are constant). Two Pfaffian systems corresponding to each one of these two cases cannot be locally isomorphic because class $(D^1 (S))$ is an invariant up to local isomorphism.

  i. class $(D^1 (S)) = 1$
By hypothesis one has $d\bar{\omega}_2 = 0 \mod (\bar{\omega}_2)$. The system $D^1 (S)$ is completely integrable; thus there is a coordinates system (y_1, y_2, y_3, y_4) such that

$$\bar{\omega}_2 = dy_2.$$

In this system the integral manifolds of $\bar{\omega}_2 = 0$ are the planes of the equation $dy_2 = 0$ and on one of these planes the form $\bar{\omega}_1$ (which satisfies $d\bar{\omega}_1 = \omega_3 \wedge \omega_4 \mod (S)$) induces a contact form. Hence the system (S) satisfies the hypothesis of the Darboux' theorem with parameters. One deduces from this theorem that (S) is locally isomorphic to the Pfaffian system

$$(S_4^5) = \begin{cases} \omega_1 = dy_1 + y_3 dy_4, \\ \omega_2 = dy_2. \end{cases}$$

ii. class $(D^1(S)) = 3$

To simplify the notations, one removes the bars of the representatives of (S). There is a basis $(\omega_1, \omega_2, \omega_3, \omega_4)$ of Pfaffian forms such that if (ω_1, ω_2) is a basis of (S) one has

$$\begin{cases} d\omega_1 = \omega_3 \wedge \omega_4 \mod (S), \\ d\omega_2 = \omega_1 \wedge \omega_3 \mod (\omega_2). \end{cases}$$

The derived system $D^1(S) = \{\omega_2\}$ being of class 3, one can choose, from Darboux' theorem, a coordinate system (y_1, y_2, y_3, y_4) of $\mathbb{R}^4$ such that ω_2 is reduced to the canonical form

$$\omega_2 = dy_2 + y_3 dy_4.$$

In this coordinate system the form ω_1 is written

$$\omega_1 = a_1 dy_1 + a_2 dy_2 + a_3 dy_3 + a_4 dy_4.$$

The change of basis $(\bar{\omega}_1 = \omega_1 - a_2 \omega_2, \bar{\omega}_2 = \omega_2)$ does not alter the written form of ω_2, and allows us to assume that a_2 is zero in the expression for ω_1. However, $\omega_1 \wedge \omega_2 \wedge d\omega_2 = 0$. One deduces that

$$(a_1 dy_1 \wedge dy_2) \wedge dy_3 \wedge dy_4 = 0,$$

yielding $a_1 = 0$, and thus $\omega_1 = a_3 dy_3 + a_4 dy_4$. As ω_1 is never zero one of the two functions a_3 or a_4 is non-zero. Notice that with the variables y_3 and y_4 playing symmetrical roles in the expression of ω_2, by replacing $y_3 dy_4$ by $y_4 dy_3 - d(y_3 y_4)$ one can assume $a_3 \neq 0$ and replace ω_1 by ω_1/a_3 what amounts writing :

$$\omega_1 = dy_3 + a_4 dy_4$$

As $\omega_1 \wedge \omega_2 \wedge d\omega_1 \neq 0$ we obtain

$$(dy_3 + a_4 dy_4) \wedge (dy_2 + y_3 dy_4) \wedge da_4 \wedge dy_4 \neq 0,$$

that is, $dy_3 \wedge dy_2 \wedge da_4 \wedge dy_4 \neq 0$. The functions (a_4, y_2, y_3, y_4) form a local coordinate system of $\mathbb{R}^4$. By setting $a_4 = \bar{y}_1$ one obtains the following local model:

$$\begin{cases} \omega_1 = dy_3 + \bar{y}_1 dy_4, \\ \omega_2 = dy_2 + y_3 dy_4. \end{cases}$$

Conclusion. Let (S) be a Pfaffian system on $\mathbb{R}^4$ of rank 2 and class 4. If the first derived system $D^1(S)$ is of rank 1 and class 3, then there is a system of local coordinates (y_1, y_2, y_3, y_4) such that the system (S) is represented by

$$(S_4^6) = \begin{cases} \omega_1 = dy_3 + y_1 dy_4, \\ \omega_2 = dy_2 + y_3 dy_4. \end{cases}$$

Theorem 5.2 *Let (S) be a Pfaffian system on $\mathbb{R}^4$ of constant rank and constant class. Let us assume that its derived systems also are of constant rank and constant class. Then (S) is locally isomorphic to the one of the following systems described in the table 2. These systems are pairwise not isomorphic.*

TABLE 2

Pfaffian systems on $\mathbb{R}^4$

	rank (S)	class (S)	rank $(D^1(S))$	class $(D^1(S))$
$S_4^1 = \begin{cases} \omega_1 = dy_1, \\ \omega_2 = dy_2 \\ \omega_3 = dy_3. \end{cases}$	3	3	3	3
$S_4^2 = \{\omega_1 = dy_1.$	1	1	1	1
$S_4^3 = \{\omega_1 = dy_1 + y_2 dy_3$	1	3	0	0
$S_4^4 = \begin{cases} \omega_1 = dy_1, \\ \omega_2 = dy_2. \end{cases}$	2	2	2	2

$S_4^5 = \begin{cases} \omega_1 = dy_1 + y_3 dy_4, \\ \omega_2 = dy_2 \end{cases}$	2	4	1	1
$S_4^6 = \begin{cases} \omega_1 = dy_3 + y_1 dy_4, \\ \omega_2 = dy_2 + y_3 dy_4 \end{cases}$	2	4	1	3
$S_4^0 = \begin{cases} \omega_1 = dy_1, \\ \omega_2 = dy_2, \\ \omega_3 = dy_3, \\ \omega_4 = dy_4 \end{cases}$	4	4	4	4

5.2.3 Examples

1. The Monge's problem

The problem consists in determining two functions $f_1(x)$ and $f_2(x)$ of a real variable satisfying a differential equation of the type

$$f_1'(x) = F\left(x, f_1(x), f_2(x), f_2'(x)\right).$$

If F is linear with respect to the variable $f_2'(x)$, then this equation has the following form

$$f_1'(x) = G(x, f_1(x), f_2(x)) + Af_2'(x).$$

One is led to find the integrals of a Pfaffian equation with 3 variables of the form

$$dz = Ady + Gdx.$$

We have

$$df_1(x) = Adf_2(x) + Gdx.$$

Set $f_2'(x) = u$ to obtain

$$df_1 = F(x, f_1, f_2, u)\, dx$$

and

$$df_2 = udx.$$

Then the problem is reduced to find the integrals of the Pfaffian system of two equations and four variables (x, f_1, f_2, u)

$$\begin{cases} \omega_1 = df_1 - Fdx \\ \omega_2 = df_2 - udx. \end{cases}$$

We have

$$\begin{cases} d\omega_1 = -dF \wedge dx \\ d\omega_2 = -du \wedge dx. \end{cases}$$

that is

$$\begin{cases} d\omega_1 = -\frac{\partial F}{\partial u} du \wedge dx - \frac{\partial F}{\partial f_1} df_1 \wedge dx - \frac{\partial F}{\partial f_2} df_2 \wedge dx \\ d\omega_2 = -du \wedge dx. \end{cases}$$

Consider the Pfaffian form

$$\bar{\omega}_1 = \omega_1 - \frac{\partial F}{\partial u} \omega_2.$$

This form verifies

$$d\bar{\omega}_1 = -\frac{\partial F}{\partial u} du \wedge dx + \frac{\partial F}{\partial u} du \wedge dx \mod (\bar{\omega}_1, \omega_2)$$

that is,

$$d\bar{\omega}_1 = 0 \mod (\bar{\omega}_1, \omega_2).$$

Thus

$$\begin{cases} (S) = (\bar{\omega}_1, \omega_2), \\ D^1 (S) = (\bar{\omega}_1). \end{cases}$$

Then the system (S) is locally isomorphic to one of the following systems (S_4^5) or (S_4^6). In order to describe the isomorphism one has to return the equation $\bar{\omega}_1 = 0$ to his canonical form.

2. Differential equations with two variables

We consider the following differential equation:

$$\frac{dz}{dx} = A\left(x, y, z, \frac{dy}{dx}\right) \frac{d^2y}{dx^2} + B\left(x, y, z, \frac{dy}{dx}\right),$$

and setting $dy/dx = u$. The above equation can be written as

$$\frac{dz}{dx} = A(x, y, z, u) \frac{du}{dx} + B(x, y, z, u).$$

This yields the Pfaffian system of two equations and for the variables

$$\begin{cases} \omega_1 = dy - u\,dx \\ \omega_2 = dz - A(x, y, z, u)\,du - B(x, y, z, u)\,dx. \end{cases}$$

3. Examples of compact manifolds provided with a Pfaffian system of type S_4^6

We will give in this paragraph some examples of 4-dimensional compact manifolds equipped with a Pfaffian system of rank 2 and class 4, whose the first derived system is of class 3 at any point.

Let n_4 be the 4–dimensional nilpotent Lie algebra whose the structural equations are :

$$\begin{cases} d\omega_1 = \omega_3 \wedge \omega_4, \\ d\omega_2 = -\omega_1 \wedge \omega_3, \\ d\omega_3 = 0, \\ d\omega_4 = 0. \end{cases}$$

This algebra admits the matricial representation

$$n_4 = \left\{ \begin{pmatrix} 0 & x_4 & x_1 & x_2 \\ 0 & 0 & x_3 & 0 \\ 0 & 0 & 0 & x_3 \\ 0 & 0 & 0 & 0 \end{pmatrix}, \ x_i \in \mathbb{R} \right\}$$

Let us denote by N_4 the connected and simply connected Lie group whose the Lie algebra is n_4. It is defined by the group of matrices

$$\begin{pmatrix} 1 & y_4 & y_1 & y_2 \\ 0 & 1 & y_3 & y_3^2/2 \\ 0 & 0 & 1 & y_3 \\ 0 & 0 & 0 & 1 \end{pmatrix}$$

with $y_i \in \mathbb{R}$. In the coordinates system (y_1, y_2, y_3, y_4), the right invariant forms ω_i on the Lie group are written

$$\begin{cases} \omega_1 = dy_1 - y_3 dy_4, \\ \omega_2 = dy_2 - y_3 dy_1 + \frac{y_3^2}{2} dy_4, \\ \omega_3 = dy_3, \\ \omega_4 = dy_4. \end{cases}$$

Let us consider the Pfaffian system (S) defined by (ω_1, ω_2) on the Lie group N_4. It is of class 4 and of rank 2, its first derived system is of class 3, and it is (globally) isomorphic to (S_4^6). The isomorphism is given by

$$\begin{cases} z_1 = y_1 + y_4, \\ z_2 = y_2 - y_1 y_3 + \frac{y_3^2 y_4}{2}, \\ z_3 = y_3 + 1, \\ z_4 = y_4. \end{cases}$$

It is clear that the Lie group N_4 is not a compact manifold. Let us consider the discrete subgroup K of N_4 formed by the matrices

$$\begin{pmatrix} 1 & n_4 & n_1 & n_2 \\ 0 & 1 & n_3 & n_3^2/2 \\ 0 & 0 & 1 & n_3 \\ 0 & 0 & 0 & 1 \end{pmatrix}$$

with $n_i \in \mathbb{Z}$. It is a closed subgroup of N_4. Thus the quotient manifold N_4/K is equipped with a structure of differential manifold of dimension 4. As one can choose a representative of each equivalence class of N_4/K with coordinates y_i less or equal to 1, the quotient manifold is compact. The Pfaffian forms ω_i being right invariant, there exist Pfaffian forms $\overline{\omega}_i$ on N_4/K such as $\omega_i = \pi^*\overline{\omega}_i$, where $\pi : N_4 \longrightarrow N_4/K$ is the canonical projection. The Pfaffian system $(\overline{S}) = (\overline{\omega}_1, \overline{\omega}_2)$ defined on N_4/K is isomorphic to (S_4^6). This describes well an example of a compact manifold equipped with such Pfaffian system.

4. Example of a compact manifold equipped with the Pfaffian system (S_4^5)

Let $\mathfrak{n}_3 \oplus \mathfrak{a}$ be the nilpotent Lie algebra of dimension 4 whose structural equations are

$$\begin{cases} d\omega_1 = \omega_3 \wedge \omega_4, \\ d\omega_2 = 0, \\ d\omega_3 = 0, \\ d\omega_4 = 0. \end{cases}$$

The connected and simply connected Lie group N corresponding to this Lie algebra is the matricial Lie group:

$$N = \left\{ \begin{pmatrix} 1 & y_3 & y_1 & y_2 \\ 0 & 1 & y_4 & 0 \\ 0 & 0 & 1 & 0 \\ 0 & 0 & 0 & 1 \end{pmatrix} \right\}.$$

In the global coordinate system (y_1, y_2, y_3, y_4), a basis of the left-invariant forms $(\omega_1, \omega_2, \omega_3, \omega_4)$ is given by

$$\begin{cases} \omega_1 = dy_1 - y_3 dy_4 \\ \omega_2 = dy_2 \\ \omega_3 = dy_3 \\ \omega_4 = dy_4. \end{cases}$$

The Pfaffian system $(S) = (\omega_1, \omega_2)$ is of type (S_4^5). Let K be the discrete subgroup formed by matrices

$$\begin{pmatrix} 1 & n_3 & n_1 & n_2 \\ 0 & 1 & n_4 & 0 \\ 0 & 0 & 1 & 0 \\ 0 & 0 & 0 & 1 \end{pmatrix}$$

with $n_i \in \mathbb{Z}$. Then N/K is a 4-dimensional compact manifold endowed with a Pfaffian system of type (S_4^5).

5.3   Pfaffian systems of five variables

Pfaffian systems of five variables has been extensively studied by Elie Cartan ([14]). In the previous paragraph, we have seen that, for dimensions ≤ 4, there exists only a finite number of models of Pfaffian systems, up to local isomorphism, the rank and the class being constant. For the 5−dimensional case, the result is quite different. We are going to put in obviousness a family of Pfaffian systems on $\mathbb{R}^5$ parametrized by a function.

5.3.1   Pfaffian systems on $\mathbb{R}^5$ of rank 1, 5 or 4

These cases are the simplest, the models are given by the Darboux and Frobenius theorems. Summarize without comments this classification.

rank$(S) = 1$

$$(S_5^1) = \{\omega_1 = dx_1.$$

$$(S_5^2) = \{\omega_1 = dx_1 + x_2 dx_3.$$

$$(S_5^3) = \{\omega_1 = dx_1 + x_2 dx_3 + x_4 dx_5.$$

rank$(S) = 5$

$$(S_5^4) = \begin{cases} \omega_1 = dx_1, \\ \omega_2 = dx_2, \\ \omega_3 = dx_3, \\ \omega_4 = dx_4, \\ \omega_5 = dx_5. \end{cases}$$

rank$(S) = 4$

$$S_5^5 = \begin{cases} \omega_1 = dx_1, \\ \omega_2 = dx_2, \\ \omega_3 = dx_3, \\ \omega_4 = dx_4. \end{cases}$$

5.3.2   Pfaffian systems of rank 3

Considering a Pfaffian system (S) on $\mathbb{R}^5$ of rank 3 at each point. Let $(\omega_1, \omega_2, \ \omega_3)$ be a basis of (S) and consider Pfaffian forms ω_4 and ω_5 such that

$$\omega_1 \wedge \omega_2 \wedge \omega_3 \wedge \omega_4 \wedge \omega_5 \neq 0.$$

We have

$$\begin{cases} d\omega_1 = A_1 \omega_4 \wedge \omega_5 \quad \mathrm{mod}\,(S), \\ d\omega_2 = A_2 \omega_4 \wedge \omega_5 \quad \mathrm{mod}\,(S), \\ d\omega_3 = A_3 \omega_4 \wedge \omega_5 \quad \mathrm{mod}\,(S). \end{cases}$$

Completely integrable case

We suppose here the three functions A_1, A_2, A_3 are identically zero. The system is completely integrable and by the Frobenius theorem, it is locally isomorphic to the following model

$$(S_5^6) = \begin{cases} \omega_1 = dx_1, \\ \omega_2 = dx_2, \\ \omega_3 = dx_3. \end{cases}$$

Non-integrable case

Assume that one of the functions A_i, for example the function A_3, is not identically zero. One works on an open subset where A_3 has no zero in order that the system is of constant class. If A_3 is without zero, the class of (S) is equal to 5. Since these does not exist a Pfaffian system of rank 3 on $\mathbb{R}^5$ of constant class 4 the class of (S) is 5. Consider the change of bases

$$\begin{cases} \overline{\omega}_1 = \omega_1 - \frac{A_1}{A_3}\omega_3, \\ \overline{\omega}_2 = \omega_2 - \frac{A_2}{A_3}\omega_3, \\ \overline{\omega}_3 = \frac{1}{A_3}\omega_3. \end{cases}$$

The forms $\bar{\omega}_1, \bar{\omega}_2, \bar{\omega}_3$ are independent at each point, and define equations of system (S). We have

$$\begin{cases} d\bar{\omega}_1 = d\omega_1 - d\left(\frac{A_1}{A_3}\right) \wedge \omega_3 - \frac{A_1}{A_3} d\omega_3 \\ d\bar{\omega}_2 = d\omega_2 - d\left(\frac{A_2}{A_3}\right) \wedge \omega_3 - \frac{A_2}{A_3} d\omega_3 \\ d\bar{\omega}_3 = \frac{1}{A_3} d\omega_3 + d\left(\frac{1}{A_3}\right) \wedge \omega_3, \end{cases}$$

that is,

$$\begin{cases} d\bar{\omega}_1 = A_1\omega_4 \wedge \omega_5 - A_1\omega_4 \wedge \omega_5 - d\left(\frac{A_1}{A_3}\right) \wedge A_3\bar{\omega}_3 \ , \ \mathrm{mod}\,(S) \\ d\bar{\omega}_2 = A_2\omega_4 \wedge \omega_5 - A_2\omega_4 \wedge \omega_5 - d\left(\frac{A_2}{A_3}\right) \wedge A_3\bar{\omega}_3, \ \mathrm{mod}\,(S) \\ d\bar{\omega}_3 = \frac{A_3}{A_3}\omega_4 \wedge \omega_5, \ \mathrm{mod}\,(S). \end{cases}$$

Thus

$$\begin{cases} d\bar{\omega}_3 = 0 \\ d\bar{\omega}_2 = 0, \ \mathrm{mod}\,(S) \\ d\bar{\omega}_3 = \omega_4 \wedge \omega_5. \end{cases}$$

Lemma 17 *Let (S) be a Pfaffian system on $\mathbb{R}^5$ of rank 3 and of constant class. If the system (S) is not completely integrable, there exists a basis $(\omega_1, \omega_2, \omega_3)$ of (S) satisfying*

$$\begin{cases} d\bar{\omega}_3 = 0 \\ d\bar{\omega}_2 = 0 \quad \mathrm{mod}\,(S), \\ d\bar{\omega}_3 = \omega_4 \wedge \omega_5, \end{cases}$$

where ω_4 and ω_5 are Pfaffian forms independent with $\omega_1, \omega_2, \omega_3$.

In this case the first derived system $D^1(S)$ of (S) is the Pfaffian system of rank 2 spanned by ω_1 and ω_2

$$D^1(S) = \{\omega_1, \omega_2\}.$$

Since $D^1(S)$ is an invariant of (S), we observe the behavior of this system, so as to deduce local models for (S).

1) $D^1(S)$ **is completely integrable**
In this case the system (S) satisfy:

$$\begin{cases} d\omega_1 = 0 & \mathrm{mod}(D^1(S)), \\ d\omega_2 = 0 & \mathrm{mod}(D^1(S)), \\ d\omega_3 = \omega_4 \wedge \omega_5 & \mathrm{mod}\,(S). \end{cases}$$

The system $D^1(S)$ being completely integrable, there exists a local coordinate system $(y_1, \cdots, y_5)$ on $\mathbb{R}^5$ such that

$$\begin{cases} \omega_1 = dy_1, \\ \omega_2 = dy_2. \end{cases}$$

Consider the (local) integral manifolds to $D^1(S)$. With respect to the coordinates system $(y_1, \cdots, y_5)$, these are defined by equations

$$\begin{aligned} y_1 &= \text{constant}, \\ y_2 &= \text{constant}. \end{aligned}$$

They are planes parallel to (y_3, y_4, y_5). Let N be an integral manifold defined by $y_1 = y_1^0$, $y_2 = y_2^0$, where y_1^0 and y_2^0 are given. Let

$$j : N \longrightarrow \mathbb{R}^5$$

be the natural injection

$$j(y_3, y_4, y_5) = \left(y_1^0, y_2^0, y_3, y_4, y_5\right).$$

Then $j^*\omega_3$ is a Pfaffian form defined on N of constant class equal to 3. In fact, we have

$$d(j^*\omega_3) = j^*\omega_4 \wedge j^*\omega_5,$$

that is a non-vanishing 2-form on N.

Now consider the situation where the derived system (S) is completely integrable. On each leaf of the integrable distribution $D^1(S)$ the form ω_3 is of constant class. According to the Darboux theorem with parameters there exists a coordinates system (x_1, x_2, x_3) on $\mathbb{R}^5$, transverse to the fibration by planes associated with the derived system (parametrized by y_1 and y_2), such that

$$\omega_3 = dx_1 + x_2 dx_3.$$

Therefore we have shown the following result :

Proposition 5.1 *Let (S) be a Pfaffian system of rank 3 of class 5, whose the first derived system is completely integrable. Then there exists a coordinates system $(x_1, x_2, x_3, x_4, x_5)$, such that (S) is represented by*

$$(S_5^7) = \begin{cases} \omega_1 = dx_1, \\ \omega_2 = dx_2, \\ \omega_3 = dx_3 + x_4 dx_5. \end{cases}$$

2) $D^1(S)$ **non-completely integrable**

Recall that (S) is represented by $(\omega_1, \omega_2, \omega_3)$, which satisfies

$$\begin{cases} d\omega_1 = 0, \\ d\omega_2 = 0 \quad \mod(S), \\ d\omega_3 = \omega_4 \wedge \omega_5 \end{cases}$$

and $D^1(S) = \{\omega_1, \omega_2\}$. Since $D^1(S)$ is not completely integrable one of functions A, B, A', B' defined by

$$\begin{cases} d\omega_1 = A\omega_3 \wedge \omega_4 + B\omega_3 \wedge \omega_5, \\ d\omega_2 = A'\omega_3 \wedge \omega_4 + B'\omega_3 \wedge \omega_5 \end{cases}$$

is not identically zero. The derived system $D^1(S)$ is assumed to be of constant rank. Therefore, we can study:
- the class of $D^1(S)$
- the second derived system $D^2(S) = D^1(D^1(S))$.

First, examine the class of $D^1(S)$.

Lemma 18 *The class of $D^1(S)$ at a point x (of the open of the definition) is equal to $c_1 = 3 + r(x)$, where $r(x)$ is the rank of the matrix*

$$M = \begin{pmatrix} A(x) & B(x) \\ A'(x) & B'(x) \end{pmatrix}.$$

Proof. By hypothesis the matrix M is not identically zero. We are going to suppose that it is of constant rank (otherwise one will be working on an open neighborhood, where the rank is constant and all our study will be valid in this open set). We have $\text{class}_x(D^1(S))$ is equal to the codimension of the following space:

$$\{X_x \in T_x\mathbb{R}^5 \setminus \omega_{ix}(X_x) = 0 \text{ and } (X \rfloor d\omega_i)_x = 0 \mod(D^1(S)), i = 1, 2\}.$$

Let X be a non-vanishing vector field verifying $\omega_1(X) = \omega_2(X) = 0$, Consider the basis $(X_1, \cdots, X_5)$ formed by vector fields, whose $(\omega_1, \cdots, \omega_5)$ is the dual basis. Therefore we have $X = a_3 X_3 + a_4 X_4 + a_5 X_5$ and

$$X \rfloor d\omega_1 = Aa_3\omega_4 - Aa_4\omega_3 + Ba_3\omega_5 - Ba_5\omega_3 \quad \mod(D^1(S))$$

$$X \rfloor d\omega_2 = A'a_3\omega_4 - A'a_4\omega_3 + B'a_3\omega_5 - B'a_5\omega_3 \quad \mod(D^1(S)).$$

Thus, $X \rfloor d\omega_1 = 0$ and $X \rfloor d\omega_2 = 0 \mod \left(D^1 \left(S \right) \right)$ imply

$$\begin{cases} Aa_3 = A'a_3 = 0, \\ Ba_3 = B'a_3 = 0, \\ Aa_4 + Ba_5 = 0, \\ A'a_4 + B'a_5 = 0. \end{cases}$$

The two last relationships can be written

$$M. \begin{pmatrix} a_4 \\ a_5 \end{pmatrix} = 0.$$

If the rank of M is constant and equal to 1 then $a_3 = 0$, and we have

$$\mathrm{co} \dim \left\{ X \backslash \omega_i \left(X \right) = 0, \, i = 1, 2 \text{ and } X \rfloor d\omega_i = 0 \mod \left(D^1 \left(S \right) \right) \right\} = 4.$$

If the rank of M is equal to 2 then $a_3 = a_4 = a_5 = 0$ and class $\left(D^1 \left(S \right) \right) = 5$. This proves the lemma. ∎

Finally, note that if the rank of M is zero, then class $\left(D^1 \left(S \right) \right) = 3$ and the system (S) is completely integrable.

This has led us to examine the two cases:

i) class $\left(D^1 \left(S \right) \right) = 4$

ii) class $\left(D^1 \left(S \right) \right) = 5$.

i) Study of the case class $\left(D^1 \left(S \right) \right) = 4$

The rank of M is constant, and it is equal to 1. By a change of basis of $D^1 \left(S \right)$ we can assume that M is constant, and we have $M = \begin{pmatrix} 0 & 0 \\ 1 & 0 \end{pmatrix}$. This is equivalent to writing:

$$\begin{cases} d\omega_1 = 0 & \mod \left(D^1 \left(S \right) \right), \\ d\omega_2 = \omega_3 \wedge \omega_4 & \mod \left(D^1 \left(S \right) \right), \\ d\omega_3 = \omega_4 \wedge \omega_5 & \mod \left(S \right). \end{cases}$$

In this case the second derived system is of constant rank equal to 1. It is defined by:

$$D^2 \left(S \right) = D^1 \left(D^1 \left(S \right) \right) = \left\{ \omega_1 \right\}.$$

Since $D^2 \left(S \right)$ is of constant rank, we can study its class and envisage the three next cases :

– class $\left(D^2 \left(S \right) \right) = 1$

– class $\left(D^2 \left(S \right) \right) = 3$

– class $(D^2(S)) = 5$.

i1) Let us consider the case class $(D^2(S)) = 1$ (and class $(D^1(S)) = 4$)

The system $D^2(S)$ is completely integrable, and in this case we are in the situation where (S) is of class 5, $D^1(S)$ is of class 4 and rank 2, $D^2(S)$ is completely integrable and it is of rank 1. Thus the Pfaffian system (S) is totally regular. The Pfaffian system $D^1(S)$ is of rank 2 and class 4, whose the first derived system is of rank 1 and completely integrable. This situation has been examined in the previous chapter; it follows from the Darboux theorem with parameters, that there exists a local coordinate system $(y_1, \cdots, y_5)$ of $\mathbb{R}^5$ such that $D^1(S)$ admits a basis (ω_1, ω_2) satisfying

$$\begin{cases} \omega_1 = dy_1, \\ \omega_2 = dy_2 + y_3 dy_4. \end{cases}$$

Consider the basis $(\omega_1, \omega_2, \omega_3)$ of (S) defined above. Since $(\omega_2, \omega_3) = D^1(S)$, we have

$$\omega_2 \wedge d\omega_2 \wedge \omega_1 \wedge \omega_3 = 0.$$

This implies

$$\begin{cases} dy_2 \wedge dy_3 \wedge dy_4 \wedge dy_1 \wedge \omega_3 = 0, \\ \omega_3 = a_1 dy_1 + a_2 dy_2 + a_3 dy_3 + a_4 dy_4. \end{cases}$$

Let $\bar{\omega}_3 = \omega_3 - a_1 \omega_1 - a_2 \omega_2$; the system (S) is represented by the forms $(\omega_1, \omega_2, \bar{\omega}_3)$, and by deleting the bars we are returned to the basis $(\omega_1, \omega_2, \omega_3)$ of (S) with

$$\begin{cases} \omega_1 = dy_1, \\ \omega_2 = dy_2 + y_3 dy_4, \\ \omega_3 = a_3 dy_3 + a_4 dy_4. \end{cases}$$

But the rank of (S) is constant and equal to 3. One of the two functions a_3 or a_4 is not identically zero. Since in the previous representation the coordinates y_3 and y_4 play symmetric roles, we can assume that a_3 is non-zero, and even without a zero. Divide ω_3 by a_3 to obtain the representation of (S) by the forms

$$\begin{cases} \omega_1 = dy_1, \\ \omega_2 = dy_2 + y_3 dy_4, \\ \omega_3 = dy_3 + a_4 dy_4. \end{cases}$$

We have

$$d\omega_3 = da_4 \wedge dy_4 = \omega_4 \wedge \omega_5 \ \mathrm{mod}\,(S).$$

Therefore

$$dy_1 \wedge dy_2 \wedge dy_3 \wedge dy_4 \wedge da_4 \neq 0$$

whose $(y_1, y_2, y_3, y_4, a_4)$ is a system of independent coordinates on $\mathbb{R}^5$. Taking $x_5 = a_4$ we obtain the following representation:

$$\begin{cases} \omega_1 = dy_1, \\ \omega_2 = dy_2 + y_3 dy_4, \\ \omega_3 = dy_3 + x_5 dy_4. \end{cases}$$

Proposition 5.2 *Let (S) be a system of rank 3 totally regular with length 2 and of class 5 such that class $\left(D^1(S)\right) = 4$. There exists a coordinates system $(y_1, y_2, y_3, y_4, y_5)$ such that (S) is represented by:*

$$(S_5^8) = \begin{cases} \omega_1 = dy_1, \\ \omega_2 = dy_2 + y_3 dy_4, \\ \omega_3 = dy_3 + y_5 dy_4. \end{cases}$$

i2) class $\left(D^2(S)\right) = 3$ (and class $\left(D^1(S)\right) = 4$)
We have the following flag system:

$$(S) \supset D^1(S) = \{\omega_1, \omega_2\} \supset D^2(S) = \{\omega_1\} \supset D^3(S) = \{0\}.$$

The first derived system satisfies

$$\begin{cases} d\omega_1 = 0 \qquad \mod \left(D^1(S)\right), \\ d\omega_2 = \omega_3 \wedge \omega_4 \mod \left(D^1(S)\right), \end{cases}$$

and $d\omega_1 \wedge \omega_1 \neq 0$. According to a suitable study in dimension 4 we are assured that there exists a coordinate system $(y_1, y_2, y_3, y_4, y_5)$ such that

$$\begin{cases} \omega_1 = dy_1 + y_2 dy_3, \\ \omega_2 = dy_2 + y_4 dy_3. \end{cases}$$

Now let us consider the basis $\{\omega_1, \omega_2, \omega_3\}$ of (S). We have

$$\omega_2 \wedge d\omega_2 \wedge \omega_1 \wedge \omega_3 = 0$$

because $\omega_2 \in D^1(S)$ thus

$$\omega_3 \wedge dy_2 \wedge dy_4 \wedge dy_3 \wedge dy_1 = 0.$$

Replacing w_3 by the form $\bar{w}_3 = w_3 - a_1 w_1 - a_2 w_2$, we can choose coefficients a_1 and a_2 such that $\bar{w}_3 = f_3 dy_3 + f_4 dy_4$. Since the rank of (S) is constant and equal to 3 one of two functions f_i is not identically zero.

α) Suppose that f_3 has no zero. We take

$$\tilde{w}_3 = \frac{\bar{w}_3}{f_3}.$$

It verifies $\bar{w}_3 = dy_3 + g dy_4$ and $d\tilde{w}_3 = dg \wedge dy_4$. The functions (y_1, y_2, y_3, y_4, g) define a local coordinates system on $\mathbb{R}^5$. By renumbering these coordinates we deduce that (S) is locally isomorphic to the Pfaffian system

$$(S_5^9) = \begin{cases} w_1 = dx_1 + x_2 dx_3, \\ w_2 = dx_2 + x_4 dx_3, \\ w_3 = dx_3 + x_5 dx_4. \end{cases}$$

β) Suppose that f_4 has no zero. The system is always regular. We take $\tilde{w}_3 = \bar{w}_3 / f_4$ and we have $\tilde{w}_3 = dy_4 + g dy_3$. As previously (y_1, y_2, y_3, y_4, g) is a local coordinates system and the Pfaffian system considered is locally isomorphic to

$$(S_5^{10}) = \begin{cases} w_1 = dx_1 + x_2 dx_3, \\ w_2 = dx_2 + x_4 dx_3, \\ w_3 = dx_4 + x_5 dx_3. \end{cases}$$

In his study of Pfaffian systems of five variables, Elie Cartan had not distinguished the two systems (S_5^9) and (S_5^{10}). Although these two systems are not locally isomorphic (see the next sections), nevertheless, they are projectively isomorphic. In the tradition of Elie Cartan, projective transformations were at the root of geometry. This probably explains the voluntary oblivion of E. Cartan, stated for the first time by Kumpera and Ruiz.

Proposition 5.3 *The systems (S_5^9) and (S_5^{10}) are not locally isomorphic.*

Proof. The distribution of planes defined by (S_5^9) (the kernel of (S_5^9)) is spanned by the vector fields

$$X_1 = \frac{\partial}{\partial x_5},$$

$$X_2 = \frac{\partial}{\partial x_4} - x_5 \frac{\partial}{\partial x_3} + x_4 x_5 \frac{\partial}{\partial x_2} + x_2 x_5 \frac{\partial}{\partial x_1},$$

while that defined by $\left(S_5^{10}\right)$ is generated by

$$X_1 = \frac{\partial}{\partial x_5},$$

$$Y_2 = \frac{\partial}{\partial x_3} - x_5\frac{\partial}{\partial x_4} - x_4\frac{\partial}{\partial x_2} - x_2\frac{\partial}{\partial x_1}.$$

We have the following relationships:

$$[X_1, Y_2] = -\frac{\partial}{\partial x_4},$$

$$[X_1, X_2] = -\frac{\partial}{\partial x_3} + x_4\frac{\partial}{\partial x_2} + x_2\frac{\partial}{\partial x_1}.$$

Set

$$Y_3 = -\frac{\partial}{\partial x_4}$$

and

$$X_3 = -\frac{\partial}{\partial x_3} + x_4\frac{\partial}{\partial x_2} + x_2\frac{\partial}{\partial x_1}.$$

The vector fields (X_1, Y_2, Y_3) generate the kernel of $D^1\left(S_5^{10}\right)$ and the vector fields (X_1, X_2, X_3) generate that of $D^1\left(S_5^9\right)$. We have

$$
\begin{aligned}
[X_1, Y_2] &= Y_3, \\
[X_1, Y_3] &= 0, \\
[Y_2, Y_3] &= Y_4 = -\frac{\partial}{\partial x_2}, \\
[Y_2, Y_4] &= -\frac{\partial}{\partial x_1} = Y_5, \\
[Y_3, Y_4] &= 0,
\end{aligned}
$$

the other undefined brackets being zero, and

$$
\begin{aligned}
[X_1, X_2] &= X_3, \\
[X_1, X_3] &= 0, \\
[X_2, X_3] &= \frac{\partial}{\partial x_2} = X_4, \\
[X_3, X_4] &= \frac{\partial}{\partial x_1} = X_5, \\
[X_2, X_4] &= x_5\frac{\partial}{\partial x_1} = x_5 X_5,
\end{aligned}
$$

where the undefined brackets are assumed to be zero. These relations prove that the vector fields $(X_1, Y_2, Y_3, Y_4, Y_5)$ generate a Lie algebra $\mathfrak{g}$ of dimension 5. Therefore the system (S_5^{10}) is associated with a linear system invariant on the Lie group corresponding to $\mathfrak{g}$. On the other hand these does not exist a Lie algebra of dimension 5 spanned by $\ker (S_5^9)$. Thus the systems (S_5^{10}) and (S_5^9) cannot be globally isomorphic. ∎

Proposition 5.4 *Let (S) be a Pfaffian system of rank 3, totally regular of length 2, of class 5 such that* class $(D^2 (S)) = 3$. *Then (S) is locally isomorphic to one of the two systems:*

$$(S_5^9) = \begin{cases} \omega_1 = dx_1 + x_2 dx_3, \\ \omega_2 = dx_2 + x_4 dx_3, \\ \omega_3 = dx_3 + x_5 dx_4. \end{cases}$$

$$(S_5^{10}) = \begin{cases} \omega_1 = dx_1 + x_2 dx_3, \\ \omega_2 = dx_2 + x_4 dx_3, \\ \omega_3 = dx_4 + x_5 dx_3. \end{cases}$$

i3) Let us consider the last case class $(D^2 (S)) = 5$ (class $(D^1 (S)) = 4$). Recall the hypothesis

$$\begin{cases} d\omega_1 = 0, \\ d\omega_2 = 0, \\ d\omega_3 = \omega_4 \wedge \omega_5 \end{cases} \mod (S),$$

and

$$\begin{cases} d\omega_1 = 0, \\ d\omega_2 = \omega_3 \wedge \omega_4, \\ d\omega_3 = \omega_4 \wedge \omega_5. \end{cases} \mod (D^1 (S)).$$

Since $D^2 (S) = \{\omega_1\}$ the Pfaffian form ω_1 must be of maximum class. But $d\omega_1 = 0 \mod (D^1 (S))$ implies $d\omega_1 = a\omega_1 \wedge \beta_1 + b\omega_2 \wedge \beta_2$ and $\omega_1 \wedge (d\omega_1)^2 = 0$. Therefore this case is excluded.

ii) Let us assume now class $(D^1 (S)) = 5$. Here we have

$$\begin{cases} d\omega_1 = A\omega_3 \wedge \omega_4 + B\omega_3 \wedge \omega_5, \\ d\omega_2 = A'\omega_3 \wedge \omega_4 + B'\omega_3 \wedge \omega_5 \end{cases} \mod (D^1 (S))$$

and the matrix

$$M = \begin{pmatrix} A & B \\ A' & B' \end{pmatrix}$$

is assumed to be of constant rank equal to 2 at each point. Therefore the second derived system $D^2(S)$ is reduced to $\{0\}$.

Proposition 5.5 *Let (S) be a Pfaffian system of rank 3 and class 5 for which first derived system is of rank 2 and class 5. Then this system is totally regular and of length 1:*

$$(S) \supset D^1(S) \supset D^2(S) = \{0\}.$$

Now determine the canonical form of such a system. Set

$$\begin{pmatrix} \overline{\omega}_1 \\ \overline{\omega}_2 \end{pmatrix} = M^{-1} \begin{pmatrix} \omega_1 \\ \omega_2 \end{pmatrix}.$$

Then $\overline{\omega}_1, \overline{\omega}_2$ and ω_3 generate (S) and satisfy

$$\begin{cases} d\overline{\omega}_1 = \omega_3 \wedge \omega_4 \mod \left(D^1(S)\right), \\ d\overline{\omega}_2 = \omega_3 \wedge \omega_5 \mod \left(D^1(S)\right), \\ d\omega_3 = \omega_4 \wedge \omega_5 \mod (S). \end{cases}$$

Similarly, we name these forms $\omega_1, \omega_2, \omega_3$ to simplify the handwriting and one considers a local coordinates system $(x_1, \cdots, x_5)$ on $\mathbb{R}^5$. Set

$$\omega_1 = a_1 dx_1 + a_2 dx_2 + a_3 dx_3 + a_4 dx_4 + a_5 dx_5.$$

We have

$$d\omega_1 = da_1 \wedge dx_1 + da_2 \wedge dx_2 + da_3 \wedge dx_3 + da_4 \wedge dx_4 + da_5 \wedge dx_5$$

and

$$\omega_1 \wedge (d\omega_1)^2 = \Delta dx_1 \wedge dx_2 \wedge dx_3 \wedge dx_4 \wedge dx_5$$

where

$$\Delta = \begin{vmatrix} \partial a_1/\partial x_1 & \partial a_1/\partial x_2 & \partial a_1/\partial x_3 & \partial a_1/\partial x_4 & \partial a_1/\partial x_5 & a_1 \\ \partial a_2/\partial x_1 & \partial a_2/\partial x_2 & \partial a_2/\partial x_3 & \partial a_2/\partial x_4 & \partial a_2/\partial x_5 & a_2 \\ \vdots & \vdots & \vdots & \vdots & \vdots & \vdots \\ \partial a_5/\partial x_1 & \partial a_5/\partial x_2 & \partial a_5/\partial x_3 & \partial a_5/\partial x_4 & \partial a_5/\partial x_5 & a_5 \\ a_1 & a_2 & a_3 & a_4 & a_5 & 0 \end{vmatrix}$$

The equation $\omega_1 = 0$ is of class 3 if and only if $\Delta = 0$.

Lemma 19 *There exists a Pfaffian form* $\bar{\omega}_1 = \lambda_1\omega_1 + \lambda_2\omega_2$ *such that the equation* $\bar{\omega}_1 = 0$ *is of class 3.*

Proof. In fact, set $\omega = \lambda_1\omega_1 + \lambda_2\omega_2$. It is not difficult to see that we have

$$
\begin{aligned}
\omega \wedge d\omega^2 \;=\; & -\lambda_1^2 \left[d\lambda_2 \wedge \omega_1 \wedge d\omega_1 \wedge \omega_2 - \lambda_2\omega_1 \wedge d\omega_1 \wedge d\omega_2 \right] \\
& + \lambda_1 \left[-\lambda_2 d\lambda_2 \wedge \omega_1 \wedge \omega_2 \wedge d\omega_2 + \lambda_2^2\omega_2 \wedge d\omega_1 \wedge d\omega_2 \right] \\
& - \lambda_1\lambda_2 \left[d\lambda_1 \wedge \omega_2 \wedge \omega_1 \wedge d\omega_1 \right] + \lambda_2^2 d\lambda_1 \wedge \omega_1 \wedge \omega_2 \wedge d\omega_2.
\end{aligned}
$$

The equation $\omega \wedge (d\omega)^2 = 0$ appears as a linear partial differential equation of the first order with respect to the function λ_1. One chooses a solution of this equation and the lemma is proved. ∎

Therefore, we can assume that the system $\{\omega_1\}$ is of class 3, and according to the Darboux' theorem we can choose a local coordinate system $(y_1, \cdots, y_5)$ such that

$$\omega_1 = dy_1 + y_3 dy_4.$$

The derived system $D^1(S)$ being spanned by ω_1 and ω_2, we can assume that with respect to the coordinate system (y_i) the form ω_2 may be written

$$\omega_2 = b_2 dy_2 + b_3 dy_3 + b_4 dy_4 + b_5 dy_5.$$

One of the two functions b_2 or b_5 is non-zero, because the class of $D^1(S)$ is maximum. One works on an open subset where one of these functions, for example b_2, has no zero and one can assume that

$$\omega_2 = dy_2 + b_3 dy_3 + b_4 dy_4 + b_5 dy_5.$$

Lemma 20 *There exists a function* h *and an integrating factor* K *such that*

$$dh = \frac{\partial h}{\partial y_1} dy_1 + \frac{\partial h}{\partial y_3} dy_3 + \frac{\partial h}{\partial y_4} dy_4 + K dy_2 + K b_5 dy_5.$$

In fact, there exists a function K such that

$$
\begin{aligned}
K(dy_2 + b_5 dy_5) \;&=\; \frac{\partial h}{\partial y_2} dy_2 + \frac{\partial h}{\partial y_5} dy_5 \\
&= dh - \frac{\partial h}{\partial y_1} dy_1 - \frac{\partial h}{\partial y_3} dy_3 - \frac{\partial h}{\partial y_4} dy_4.
\end{aligned}
$$

Take for h a function independent of the variable y_1. This allows us to return the written form of ω_2 to

$$\omega_2 = dy_2 + b_3 dy_3 + b_4 dy_4.$$

We have

$$\begin{cases} d\omega_2 = db_3 \wedge dy_3 + db_4 \wedge dy_4, \\ d\omega_1 = dy_3 \wedge dy_4. \end{cases}$$

Since $D^1(S)$ is of class 5 one of the two systems

$$(y_1, y_3, y_4, y_2, b_3) \quad \text{or} \quad (y_1, y_2, y_3, y_4, b_4)$$

is a local coordinate system. Since y_3 and y_4 play symmetric roles we can assume that $dy_1 \wedge dy_2 \wedge dy_3 \wedge dy_4 \wedge db_3 \neq 0$. Therefore one has proved:

Proposition 5.6 *Under the previous hypothesis $D^1(S)$ is locally isomorphic to*

$$\begin{cases} \omega_1 = dy_1 + y_3 dy_4, \\ \omega_2 = dy_2 + y_5 dy_3 + f dy_4. \end{cases}$$

Now consider the Pfaffian form ω_3. It is of the form

$$\omega_3 = c_3 dy_3 + c_4 dy_4 + c_5 dy_5.$$

Since we have

$$\begin{cases} d\omega_1 = dy_3 \wedge dy_4 = \omega_3 \wedge \omega_4 \quad \mod (\omega_1, \omega_2), \\ d\omega_2 = dy_5 \wedge dy_3 + df \wedge dy_4 = \omega_3 \wedge \omega_5 \quad \mod (\omega_1, \omega_2), \end{cases}$$

one of functions c_3 or c_4 is non-zero. Suppose $c_3 \neq 0$ and

$$\omega_3 = dy_3 + c_4 dy_4 + c_5 dy_5.$$

The identity $dy_3 \wedge dy_4 = \omega_3 \wedge \omega_4 \mod (D^1(S))$ implies $c_5 = 0$ then

$$\omega_3 = dy_3 + c_4 dy_4.$$

Similarly

$$d\omega_2 = dy_5 \wedge dy_3 + df \wedge dy_4 = \omega_3 \wedge \omega_5 \mod (\omega_1, \omega_2)$$

implies

$$dy_5 \wedge dy_3 + df \wedge dy_4 = dy_3 \wedge \omega_5 + c_4 dy_4 \wedge \omega_5 \mod (\omega_1, \omega_2)$$

and

$$c_4 = -\frac{\partial f}{\partial y_5}.$$

Now if we assume that $c_4 \neq 0$ thus we can writte $w_3 = dy_4 + c_3 dy_3 + c_5 dy_5$. We also obtain

$$c_5 = 0 \quad \text{and} \quad c_3 = -\left(\frac{\partial f}{\partial y_5}\right)^{-1}.$$

One finds the previous case as soon as $\partial f / \partial y_5 \neq 0$.

Proposition 5.7 Let (S) be a Pfaffian system of rank 3, totally regular of length 1, of class 5 such that class $\left(D^1(S)\right) = 5$. Then (S) is locally isomorphic to the one of systems:

$$\left(S_5^{11}(f)\right) = \begin{cases} w_1 = dy_1 + y_3 dy_4, \\ w_2 = dy_2 + y_5 dy_3 + f dy_4, \\ w_3 = dy_3 - \frac{\partial f}{\partial y_5} dy_4. \end{cases}$$

The derived system is of class 5 as soon as f is not a linear function on y_5. This implies that $\partial f / \partial y_5^2$ is non zero. Take $\partial f / \partial y_5$ as a new variable y_5. The equation $w_3 = 0$ give $dy_3 = y_5 dy_4$ and the system may be written

$$\left(S_5^{11}(F)\right) = \begin{cases} w_1 = dy_1 + y_3 dy_4, \\ w_2 = dy_2 + F dy_4, \\ w_3 = dy_3 - y_5 dy_4 \end{cases}$$

where $F = f + y_5^2$. Note that this written form does not respect the flag decomposition of (S), the derived system is generated by the forms w_1 and $w_2 + (\partial f / \partial y_5) w_3$.

Remark 9 One does not determine in this chapter the relationships concerning the parameter f in order that two systems $S_5^{11}(f)$ and $S_5^{11}(f')$ are locally isomorphic. Nevertheless, one may be assured that there exists an infinity of models, up to local isomorphism, of Pfaffian systems of five variables (which was not the case in $\mathbb{R}^4$).

5.3.3   Pfaffian systems of rank 2

We consider now a Pfaffian system (S) of constant rank equal to 2 on $\mathbb{R}^5$. Let w_1 and w_2 be representatives of (S). By choosing a frame $(w_1, w_2, w_3, w_4, w_5)$ of Pfaffian forms, we have

$$\begin{cases} dw_1 = A_1 w_3 \wedge w_4 + A_2 w_3 \wedge w_5 + A_3 w_4 \wedge w_5 \quad \text{mod}(S), \\ dw_2 = B_1 w_3 \wedge w_4 + B_2 w_3 \wedge w_5 + B_3 w_4 \wedge w_5 \quad \text{mod}(S). \end{cases}$$

class $(S) = 2$

The system (S) is completely integrable. It is locally isomorphic to

$$(S_5^{12}) = \begin{cases} \omega_1 = dx_1, \\ \omega_2 = dx_2. \end{cases}$$

class $(S) = 3$

Consider the matrix

$$M = \begin{pmatrix} A_1 & A_2 & A_3 \\ B_1 & B_2 & B_3 \end{pmatrix}.$$

Since (S) is not completely integrable the matrix M is not identically zero. The class of (S) being constant, the rank of M is constant. If the rank of M is equal to 1 then class $(S) = 4$. In fact, according to reduction we can assume that

$$M = \begin{pmatrix} 1 & 0 & 0 \\ 0 & 0 & 0 \end{pmatrix}.$$

Then the characteristic system of (S) contains the equations $\omega_3 = 0$, $\omega_4 = 0$, and we have class $(S) = 4$. If the rank of M is equal to 2 the system (S) is of maximum class. Thus the case class $(S) = 3$ is excluded.

class $(S) = 4$

The matrix M can be reduced to

$$M = \begin{pmatrix} 1 & 0 & 0 \\ 0 & 0 & 0 \end{pmatrix}$$

and the system (S) satisfies

$$\begin{cases} d\omega_1 = \omega_3 \wedge \omega_4, \\ d\omega_2 = 0 \end{cases} \mod(S),$$

which can be returned to a system on $\mathbb{R}^4$. The suitable study in the preceding paragraph shows that (S) is locally isomorphic to one of systems

$$(S_5^{13}) = \begin{cases} \omega_1 = dy_1 + y_3 dy_4, \\ \omega_2 = dy_2, \end{cases}$$

$$(S_5^{14}) \begin{cases} \omega_1 = dy_1 + y_3 dy_4, \\ \omega_2 = dy_2 + y_1 dy_4. \end{cases}$$

class $(S) = 5$

The matrix M can be reduced to

$$M = \begin{pmatrix} 1 & 0 & 0 \\ 0 & 1 & 0 \end{pmatrix}$$

and the system (S) satisfies

$$\begin{cases} d\omega_1 = \omega_3 \wedge \omega_4 \\ d\omega_2 = \omega_3 \wedge \omega_5 \end{cases} \mod (S).$$

This study has led to the preceding section. The system is isomorphic to

$$(S_5^{16}(f)) = \begin{cases} \omega_1 = dx_1 + x_3 dx_4, \\ \omega_2 = dx_2 + x_5 dx_3 + f dx_4. \end{cases}$$

This ends the classification of Pfaffian systems of 5 variables, up to the remark that the parameter is probably not a parameter separating non-isomorphic systems.

Theorem 5.3 *All Pfaffian system of five variables of constant rank and class is locally isomorphic to one of systems given in the table 3.*

TABLE 3

PFAFFIAN SYSTEMS OF 5 VARIABLES

	rank (S)	class (S)
$S_5^1 = \{\omega_1 = dx_1,$	1	1
$S_5^2 = \{\omega_1 = dx_1 + x_2 dx_3,$	1	3
$S_5^3 = \{\omega_1 = dx_1 + x_2 dx_3 + x_4 dx_5$	1	5

	rank (S)	class (S)
$S_5^4 = \begin{cases} \omega_1 = dx_1, \\ \omega_2 = dx_2, \\ \omega_3 = dx_3, \\ \omega_4 = dx_4, \\ \omega_5 = dx_5, \end{cases}$	5	5
$S_5^5 = \begin{cases} \omega_1 = dx_1, \\ \omega_2 = dx_2, \\ \omega_3 = dx_3, \\ \omega_4 = dx_4, \end{cases}$	4	4
$S_5^6 = \begin{cases} \omega_1 = dx_1, \\ \omega_2 = dx_2, \\ \omega_3 = dx_3, \end{cases}$	3	3
$S_5^7 = \begin{cases} \omega_1 = dx_1 \\ \omega_2 = dx_2 \\ \omega_3 = dx_3 + x_4 dx_5 \end{cases}$	3	5
$S_5^8 = \begin{cases} \omega_1 = dx_1, \\ \omega_2 = dx_2 + x_3 dx_4, \\ \omega_3 = dx_3 + x_5 dx_4, \end{cases}$	3	5
$S_5^9 = \begin{cases} \omega_1 = dx_1 + x_2 dx_3, \\ \omega_2 = dx_2 + x_3 dx_4, \\ \omega_3 = dx_3 + x_5 dx_4, \end{cases}$	3	5
$S_5^{10} = \begin{cases} \omega_1 = dx_1 + x_2 dx_3, \\ \omega_2 = dx_2 + x_4 dx_3, \\ \omega_3 = dx_3 + x_5 dx_3, \end{cases}$	3	5
$S_5^{11}(f) = \begin{cases} \omega_1 = dx_1 + x_3 dx_4, \\ \omega_2 = dx_2 + x_5 dx_3 + f dx_4, \\ \omega_3 = dx_3 - \frac{\partial f}{\partial x_5} dx_4, \end{cases}$	3	5
$S_5^{12} = \begin{cases} \omega_1 = dx_1, \\ \omega_2 = dx_2, \end{cases}$	2	2

		rank (S)	class (S)
$S_5^{13} = \begin{cases} \end{cases}$	$\begin{aligned} \omega_1 &= dx_1 + x_3 dx_4, \\ \omega_2 &= dx_2, \end{aligned}$	2	2
$S_5^{14} = \begin{cases} \end{cases}$	$\begin{aligned} \omega_1 &= dx_1 + x_3 dx_4, \\ \omega_2 &= dx_2 + x_1 dx_4, \end{aligned}$	2	4
$S_5^{15} = \begin{cases} \end{cases}$	$\begin{aligned} \omega_1 &= dx_1 + x_4 dx_3, \\ \omega_2 &= dx_2 + x_5 dx_3, \end{aligned}$	2	5
$S_5^{16} = \begin{cases} \end{cases}$	$\begin{aligned} \omega_1 &= dx_1 + x_3 dx_4, \\ \omega_2 &= dx_2 + x_5 dx_3 + f dx_4, \end{aligned}$	2	5

5.4 Symmetries of the systems $S_5^{11}(f)$

Consider a Pfaffian system (S) of rank 3 and class 5 whose derived system is of rank 2 and class 5. It verifies

$$
\begin{cases}
d\omega_1 = \omega_3 \wedge \omega_4 & \mod (D(S)), \\
d\omega_2 = \omega_3 \wedge \omega_5 & \mod (D(S)), \\
d\omega_3 = \omega_4 \wedge \omega_5 & \mod (D(S))
\end{cases}
$$

where $(S) = \{\omega_1, \omega_2, \omega_3\}$ and $D(S) = \{\omega_1, \omega_2\}$. Let G be the Lie subgroup of $Gl(5, \mathbb{R})$ leaving invariant (S), $D(S)$ and the previous reduction. This group is formed by the invertible matrices

$$
\begin{pmatrix}
\delta u_1 & \delta u_2 & 0 & 0 & 0 \\
\delta u_3 & \delta u_4 & 0 & 0 & 0 \\
u_5 & u_6 & \delta & 0 & 0 \\
a & b & c & u_1 & u_2 \\
d & e & f & u_3 & u_4
\end{pmatrix},
$$

where $\delta = u_1 u_4 - u_2 u_3 \neq 0$. The Lie algebra of G is given by

$$
\begin{pmatrix}
2e_1 + e_4 & e_2 & 0 & 0 & 0 \\
e_3 & e_1 + 2e_4 & 0 & 0 & 0 \\
f_1 & f_2 & e_1 + e_4 & 0 & 0 \\
f_3 & f_4 & f_5 & e_1 & e_2 \\
f_6 & f_7 & f_8 & e_3 & e_4
\end{pmatrix}.
$$

By a change of basis given by G, we can find a frame $(\alpha_1, \alpha_2, \alpha_3, \alpha_4, \alpha_5)$ such that $(S) = (\alpha_1, \alpha_2, \alpha_3)$ and $D(S) = (\alpha_1, \alpha_2)$ with

$$
\begin{aligned}
d\alpha_1 &= \alpha_1 \wedge (2\pi_1 + \pi_4) + \alpha_2 \wedge \pi_2 + \alpha_3 \wedge \alpha_4, \\
d\alpha_2 &= \alpha_1 \wedge \pi_3 + \alpha_2 \wedge (\pi_1 + 2\pi_4) + \alpha_3 \wedge \alpha_5, \\
d\alpha_3 &= \alpha_1 \wedge \pi_5 + \alpha_2 \wedge \pi_6 + \alpha_3 \wedge (\pi_1 + \pi_4) + \alpha_4 \wedge \alpha_5, \\
d\alpha_4 &= \alpha_1 \wedge \pi_7 + \frac{4}{3}\alpha_3 \wedge \pi_6 + \alpha_4 \wedge \pi_1 + \alpha_5 \wedge \pi_2, \\
d\alpha_5 &= \alpha_2 \wedge \pi_7 - \frac{4}{3}\alpha_3 \wedge \pi_5 + \alpha_4 \wedge \pi_3 + \alpha_5 \wedge \pi_4,
\end{aligned}
$$

where (π_i) are Pfaffian forms depending on α_j and G (calculations of reduction can be found in [14]). The Lie algebra $\mathfrak{g}_7$ of the subgroup of G leaving these formulae invariant is given by

$$
\mathfrak{g}_7 = \left\{ \begin{pmatrix}
2e_1 + e_4 & e_2 & 0 & 0 & 0 \\
e_3 & e_1 + 2e_4 & 0 & 0 & 0 \\
e_5 & e_6 & e_1 + e_4 & 0 & 0 \\
e_7 & 0 & \frac{4}{3}e_6 & e_1 & e_2 \\
0 & e_7 & -\frac{4}{3}e_5 & e_3 & e_4
\end{pmatrix} \right\}.
$$

The fundamental identities $dd\alpha_i = 0$ imply that

$$
\begin{aligned}
&2d\pi_1 \wedge \alpha_1 + d\pi_4 \wedge \alpha_1 + d\pi_2 \wedge \alpha_2 = \\
&\alpha_2 \wedge \pi_2 \wedge (2\pi_1 + \pi_4) + \alpha_3 \wedge \alpha_4 \wedge (2\pi_1 + \pi_4) + \alpha_1 \wedge \pi_3 \wedge \pi_2 \\
&+\alpha_2 \wedge (\pi_1 + 2\pi_4) \wedge \pi_2 + \alpha_3 \wedge \alpha_5 \wedge \pi_2 + \alpha_1 \wedge \pi_5 \wedge \alpha_4 + \alpha_2 \wedge \pi_6 \wedge \alpha_4 \\
&+\alpha_3 \wedge (\pi_1 + \pi_4) \wedge \alpha_4 - \alpha_3 \wedge \alpha_1 \wedge \pi_7 - \alpha_3 \wedge \alpha_4 \wedge \pi_1 \\
&-\alpha_3 \wedge \alpha_5 \wedge \pi_2,
\end{aligned}
$$

$$
\begin{aligned}
&d\pi_3 \wedge \alpha_1 + d\pi_1 \wedge \alpha_2 + 2d\pi_4 \wedge \alpha_2 = \\
&\alpha_1 \wedge (2\pi_1 + \pi_4) \wedge \pi_3 + \alpha_2 \wedge \pi_2 \wedge \pi_3 + \alpha_3 \wedge \alpha_4 \wedge \pi_3 + \alpha_2 \wedge \pi_6 \wedge \alpha_5 \\
&+\alpha_1 \wedge \pi_3 \wedge (\pi_1 + 2\pi_4) + \alpha_3 \wedge \alpha_5 \wedge (\pi_1 + 2\pi_4) + \alpha_1 \wedge \pi_5 \wedge \alpha_5 \\
&+\alpha_3 \wedge (\pi_1 + \pi_4) \wedge \alpha_5 - \alpha_3 \wedge \alpha_2 \wedge \pi_7 \\
&-\alpha_3 \wedge \alpha_4 \wedge \pi_3 - \alpha_3 \wedge \alpha_5 \wedge \pi_4,
\end{aligned}
$$

$$
\begin{aligned}
&d\pi_5 \wedge \alpha_1 + d\pi_6 \wedge \alpha_2 \wedge + d\pi_1 \wedge \alpha_3 + d\pi_4 \wedge \alpha_3 = \\
&\alpha_1 \wedge (2\pi_1 + \pi_4) \wedge \pi_5 + \alpha_2 \wedge \pi_2 \wedge \pi_5 + \alpha_3 \wedge \alpha_4 \wedge \pi_5 + \alpha_1 \wedge \pi_3 \wedge \pi_6 \\
&+\alpha_3 \wedge \alpha_5 \wedge \pi_6 + \alpha_2 \wedge (\pi_1 + 2\pi_4) \wedge \pi_6 + \alpha_1 \wedge \pi_5 \wedge (\pi_1 + \pi_4) \\
&+\alpha_2 \wedge \pi_6 \wedge (\pi_1 + \pi_4) + \alpha_4 \wedge \alpha_5 \wedge (\pi_1 + \pi_4) \\
&+\alpha_1 \wedge \pi_7 \wedge \alpha_5 + \frac{4}{3}\alpha_3 \wedge \pi_6 \wedge \alpha_5 + \alpha_4 \wedge \pi_1 \wedge \alpha_5 - \alpha_4 \wedge \alpha_2 \wedge \pi_7 \\
&+\alpha_4 \wedge \frac{4}{3}\alpha_3 \wedge \pi_5 - \alpha_4 \wedge \alpha_5 \wedge \pi_4,
\end{aligned}
$$

$$d\pi_7 \wedge \alpha_1 + d\pi_6 \wedge \tfrac{4}{3}\alpha_3 + d\pi_1 \wedge \alpha_4 + d\pi_2 \wedge \alpha_5 =$$
$$\alpha_1 \wedge (2\pi_1 + \pi_4) \wedge \pi_7 + \alpha_2 \wedge \pi_2 \wedge \pi_7 + \alpha_3 \wedge \alpha_4 \wedge \pi_7 + \tfrac{4}{3}\alpha_1 \wedge \pi_5 \wedge \pi_6$$
$$+\alpha_3 \wedge (\pi_1 + \pi_4) \wedge \pi_6 + \alpha_4 \wedge \alpha_5 \wedge \pi_6) + \alpha_1 \wedge \pi_7 \wedge \pi_1$$
$$+\tfrac{4}{3}\alpha_3 \wedge \pi_6 \wedge \pi_1 + \alpha_5 \wedge \pi_2 \wedge \pi_1 + \alpha_2 \wedge \pi_7 \wedge \pi_2 - \tfrac{4}{3}\alpha_3 \wedge \pi_5 \wedge \pi_2$$
$$+\alpha_4 \wedge \pi_3 \wedge \pi_2 + \alpha_5 \wedge \pi_4 \wedge \pi_2.$$

From this system we deduce the expressions of $d\pi_i$:

$$
\begin{aligned}
d\pi_1 \;=\;& \pi_3 \wedge \pi_2 + \frac{1}{3}\alpha_3 \wedge \pi_7 - \frac{2}{3}\alpha_4 \wedge \pi_5 + \frac{1}{3}\alpha_5 \wedge \pi_6 + \alpha_1 \wedge \chi_1 \\
& + 2B_2\alpha_1 \wedge \alpha_3 + B_3\alpha_2 \wedge \alpha_3 + 2A_2\alpha_1 \wedge \alpha_4 + 2A_3\alpha_1 \wedge \alpha_5 \\
& + A_3\alpha_2 \wedge \alpha_4 + A_4\alpha_2 \wedge \alpha_5, \\
d\pi_2 \;=\;& \pi_2 \wedge (\pi_1 - \pi_4) - \alpha_4 \wedge \pi_6 + \alpha_1 \wedge \chi_2 + B_4\alpha_2 \wedge \alpha_3 + A_4\alpha_2 \wedge \alpha_4 \\
& + A_5\alpha_2 \wedge \alpha_5, \\
d\pi_3 \;=\;& \pi_4 \wedge (\pi_4 - \pi_1) - \alpha_5 \wedge \pi_5 + \alpha_2 \wedge \chi_1 - B_1\alpha_1 \wedge \alpha_3 - A_1\alpha_1 \wedge \alpha_4 \\
& - A_2\alpha_1 \wedge \alpha_5, \\
d\pi_4 \;=\;& \pi_2 \wedge \pi_3 + \frac{1}{3}\alpha_3 \wedge \pi_7 + \frac{1}{3}\alpha_4 \wedge \pi_5 - \frac{2}{3}\alpha_5 \wedge \pi_6 + \alpha_2 \wedge \chi_2 \\
& - B_2\alpha_1 \wedge \alpha_3 - 2B_3\alpha_2 \wedge \alpha_3 - A_2\alpha_1 \wedge \alpha_4 - A_3\alpha_1 \wedge \alpha_5 \\
& - 2A_3\alpha_2 \wedge \alpha_4 - 2A_4\alpha_2 \wedge \alpha_5, \\
d\pi_5 \;=\;& \pi_1 \wedge \pi_5 + \pi_3 \wedge \pi_6 - \alpha_5 \wedge \pi_7 + \alpha_3 \wedge \chi_1 - \frac{9}{32}D_2\alpha_1 \wedge \alpha_2 \\
& + \frac{9}{8}C_1\alpha_1 \wedge \alpha_3 + \frac{9}{8}C_2\alpha_2 \wedge \alpha_3 + A_2\alpha_3 \wedge \alpha_4 + A_3\alpha_3 \wedge \alpha_5 \\
& + \frac{3}{4}B_1\alpha_1 \wedge \alpha_4 + \frac{3}{4}B_2(\alpha_1 \wedge \alpha_5 + \alpha_2 \wedge \alpha_4) + \frac{3}{4}B_3\alpha_2 \wedge \alpha_5, \\
d\pi_6 \;=\;& \pi_2 \wedge \pi_5 + \pi_4 \wedge \pi_6 + \alpha_4 \wedge \pi_7 + \alpha_3 \wedge \chi_2 + \frac{9}{32}D_2\alpha_1 \wedge \alpha_2 \\
& + \frac{9}{8}C_1\alpha_1 \wedge \alpha_3 + \frac{9}{8}C_3\alpha_2 \wedge \alpha_3 - A_3\alpha_3 \wedge \alpha_4 - A_4\alpha_3 \wedge \alpha_5 \\
& + \frac{3}{4}B_2\alpha_1 \wedge \alpha_4 + \frac{3}{4}B_3(\alpha_1 \wedge \alpha_5 + \alpha_2 \wedge \alpha_4) + \frac{3}{4}B_4\alpha_2 \wedge \alpha_5, \\
d\pi_7 \;=\;& \frac{4}{3}\pi_3 \wedge \pi_6 + (\pi_1 + \pi_4) \wedge \pi_7 + \alpha_4 \wedge \chi_1 + \alpha_5 \wedge \chi_2 + \frac{9}{64}E\alpha_1 \wedge \alpha_2 \\
& - \frac{3}{8}D_1\alpha_1 \wedge \alpha_3 - \frac{3}{8}D_2\alpha_2 \wedge \alpha_3 + 2A_3\alpha_4 \wedge \alpha_5 - B_2\alpha_3 \wedge \alpha_4 \\
& + B_3\alpha_3 \wedge \alpha_5.
\end{aligned}
$$

A change of basis defined by the Lie group G_7, whose the Lie algebra is $\mathfrak{g}_7$, transforms the coefficients (A_i, B_i, C_i, D_i, E) into $(A'_i, B'_i, C'_i, D'_i, E')$, the transformations formulae correspond to an action of a Lie group whose the Lie algebra is the Lie algebra of matrices given by:

$$
\begin{aligned}
A'_1 &= -4e_1 A_1 - 4e_3 A_2, \\
A'_2 &= -e_2 A_1 - (3e_1 + e_4) A_2 - 3e_3 A_3, \\
A'_3 &= -2e_2 A_1 - (2e_1 + 2e_4) A_3 - 2e_3 A_4, \\
A'_4 &= -3e_2 A_3 - (e_1 + 3e_4) A_4 - e_3 A_5, \\
A'_5 &= -4e_2 A_4 - 4e_4 A_5, \\
B'_1 &= -\frac{4}{3} e_6 A_1 + \frac{4}{3} e_5 A_2 - (4e_1 + e_4) B_1 - 3e_3 B_2, \\
B'_2 &= -\frac{4}{3} e_6 A_2 + \frac{4}{3} e_5 A_3 - e_2 B_1 - (3e_1 + 2e_4) B_2 - 2e_3 B_3, \\
B'_3 &= -\frac{4}{3} e_6 A_3 + \frac{4}{3} e_5 A_4 - 2e_2 B_2 - (2e_1 + 3e_4) B_3 - e_3 B_4, \\
B'_4 &= -\frac{4}{3} e_6 A_4 + \frac{4}{3} e_5 A_5 - 3e_2 B_3 - (e_1 + 4e_4) B_4, \\
C'_1 &= -\frac{8}{3} e_6 B_1 + \frac{8}{3} e_5 B_2 - 2(2e_1 + e_4) C_1 - 2e_3 C_2, \\
C'_2 &= -\frac{8}{3} e_6 B_2 + \frac{8}{3} e_5 B_3 - e_2 C_1 - 3(e_1 + e_4) C_2 - e_3 C_3, \\
C'_3 &= -\frac{8}{3} e_6 B_3 + \frac{8}{3} e_5 B_4 - 2e_2 C_2 - 2(e_1 + 2e_4) C_3, \\
D'_1 &= -4e_6 C_1 + 4e_5 C_2 - (4e_1 + 3e_4) D_1 - e_3 D_2, \\
D'_2 &= -4e_6 C_2 + 4e_5 C_3 - e_2 D_1 - (3e_1 + 4e_4) D_2, \\
E' &= \frac{16}{3} e_6 D_1 - \frac{16}{3} e_5 D_2 - 4(e_1 + e_4) E.
\end{aligned}
$$

The Lie subalgebra formed by the matrices

$$
\begin{pmatrix}
-4e_1 & -4e_3 & & & \\
-e_2 & -3e_1 - e_4 & -3e_3 & & \\
& -2e_2 & -2e_1 - 2e_4 & -2e_3 & \\
& & -3e_2 & -e_1 - 3e_4 & -e_3 \\
& & & -4e_2 & -4e_4
\end{pmatrix}
$$

leaves A_i invariant. Equally the previous formulae show that there exists a subalgebra corresponding to transformations of A_i and B_i.

Theorem 5.4 *Let*

$$
\mathcal{F}(\alpha_4, \alpha_5) = A_1 \alpha_4^4 + 4A_2 \alpha_4^3 \alpha_5 + 6A_3 \alpha_4^2 \alpha_5^2 + 4A_4 \alpha_4 \alpha_5^3 + A_5 \alpha_5^4
$$

and

$$\begin{aligned}
\mathcal{G}(\alpha_3, \alpha_4, \alpha_5) = {} & A_1\alpha_4^4 + 4A_2\alpha_4^3\alpha_5 + 6A_3\alpha_4^2\alpha_5^2 + 4A_4\alpha_4\alpha_5^3 + A_5\alpha_5^4 \\
& + 4(B_1\alpha_4^3 + 3B_2\alpha_4^2\alpha_5 + 3B_3\alpha_4\alpha_5^2 + B_4\alpha_5^3)\alpha_3 \\
& + 6(C_1\alpha_4^2 + 2C_2\alpha_4\alpha_5 + C_2\alpha_5^2)\alpha_3^2 \\
& + 4(D_1\alpha_4 + D_2\alpha_5)\alpha_3^3 + E\alpha_3^4
\end{aligned}$$

be 4-forms. If two Pfaffian systems $(S) = (\alpha_1, \alpha_2, \alpha_3, \alpha_4, \alpha_5)$ *and* $(S') = (\alpha_1', \alpha_2', \alpha_3', \alpha_4', \alpha_5')$ *of type* $S_5^{11}(f)$ *are equivalent, then the corresponding* $\mathcal{F}(\alpha_4, \alpha_5)$ *and* $\mathcal{F}(\alpha_4', \alpha_5')$ *(respectively* $\mathcal{G}(\alpha_3, \alpha_4, \alpha_5)$ *and* $\mathcal{G}(\alpha_3', \alpha_4', \alpha_5'))$ *are deduced from each other via the previous transformations formulae .*

Proof. In fact,

$$\begin{aligned}
A_1' &= -4e_1 A_1 - 4e_3 A_2, \\
A_2' &= -e_2 A_1 - (3e_1 + e_4) A_2 - 3e_3 A_3, \\
A_3' &= -2e_2 A_1 - (2e_1 + 2e_4) A_3 - 2e_3 A_4, \\
A_4' &= -3e_2 A_3 - (e_1 + 3e_4) A_4 - e_3 A_5, \\
A_5' &= -4e_2 A_4 - 4e_4 A_5,
\end{aligned}$$

thus,

$$\begin{aligned}
\mathcal{F}(\alpha_4', \alpha_5') = {} & A_1'\alpha_4'^4 + 4A_2'\alpha_4'^3\alpha_5' + 6A_3'\alpha_4'^2\alpha_5'^2 + 4A_4'\alpha_4'\alpha_5'^3 + A_5'\alpha_5'^4 \\
& + 4A_2(-e_3\alpha_4'^4 - (3e_1 + e_4)\alpha_4'^3\alpha_5') \\
& + 6A_3(-2e_3\alpha_4'^3\alpha_5' - (2e_1 + 2e_4)\alpha_4'^2\alpha_5'^2 - 2e_2\alpha_4'\alpha_5'^3) \\
& + 4A_4(-3e_3\alpha_4'^2\alpha_5'^2 - (e_1 + 3e_4)\alpha_4'\alpha_5'^3 - e_2\alpha_5'^4) \\
& + A_5(-e_3\alpha_4'\alpha_5'^3 - 4e_4\alpha_5'^4).
\end{aligned}$$

Since

$$\left\{ \begin{aligned}
d\alpha_1 &= \alpha_1 \wedge (2\pi_1 + \pi_4) + \alpha_2 \wedge \pi_2 + \alpha_3 \wedge \alpha_4, \\
d\alpha_2 &= \alpha_1 \wedge \pi_3 + \alpha_2 \wedge (\pi_1 + 2\pi_4) + \alpha_3 \wedge \alpha_5, \\
d\alpha_3 &= \alpha_1 \wedge \pi_5 + \alpha_2 \wedge \pi_6 + \alpha_3 \wedge (\pi_1 + \pi_4) + \alpha_4 \wedge \alpha_5, \\
d\alpha_4 &= \alpha_1 \wedge \pi_7 + \tfrac{4}{3}\alpha_3 \wedge \pi_6 + \alpha_4 \wedge \pi_1 + \alpha_5 \wedge \pi_2, \\
d\alpha_5 &= \alpha_2 \wedge \pi_7 - \tfrac{4}{3}\alpha_3 \wedge \pi_5 + \alpha_4 \wedge \pi_3 + \alpha_5 \wedge \pi_4,
\end{aligned} \right.$$

we deduce the variation of the binary form $\mathcal{F}(\alpha_4, \alpha_5)$

$$\begin{aligned}
& 4A_1\alpha_4^3 d\alpha_4 + 12A_2\alpha_4^2\alpha_5 d\alpha_4 + 4A_2\alpha_4^3 d\alpha_5 + 12A_3\alpha_4\alpha_5^2 d\alpha_4 \\
& + 12A_3\alpha_4^2\alpha_5 d\alpha_5 + + 4A_4\alpha_5^3 d\alpha_4 + 12A_4\alpha_4\alpha_5^2 d\alpha_5 + 4A_5\alpha_5^3 d\alpha_5 =
\end{aligned}$$

$$4A_1\alpha_4^3(e_1\alpha_4 + e_2\alpha_5) + 12A_2\alpha_4^2\alpha_5(e_1\alpha_4 + e_2\alpha_5) + 4A_2\alpha_4^3(e_3\alpha_4 + e_4\alpha_5)$$
$$+12A_3\alpha_4\alpha_5^2(e_1\alpha_4 + e_2\alpha_5) + 12A_3\alpha_4^2\alpha_5(e_3\alpha_4 + e_4\alpha_5)$$
$$+4A_4\alpha_5^3(e_1\alpha_4 + e_2\alpha_5) + 12A_4\alpha_4\alpha_5^2(e_3\alpha_4 + e_4\alpha_5) + 4A_5\alpha_5^3(e_3\alpha_4 + e_4\alpha_5)$$

$$=$$

$$\alpha_4^4(4e_1A_1 + 4e_3A_2) + 4\alpha_4^3\alpha_5(e_2A_1 + (3e_1 + e_4)A_2 + 3e_3A_3)$$
$$+6\alpha_4^2\alpha_5^2(2e_2A_1 + (2e_1 + 2e_4)A_3 + 2e_3A_4) + 4\alpha_4\alpha_5^3(3e_2A_3$$
$$+(e_1 + 3e_4)A_4 + e_3A_5) + \alpha_5^4(4e_2A_4 + 4e_4A_5)   \quad \mathrm{mod}(\alpha_1, \alpha_2, \alpha_3).$$

This formula corresponds to the binary form associated with the system S'. The calculation is identical concerning the ternary form $\mathcal{G}$. This proves the theorem. ■

5.4.1   Application: Reduction of $S_5^{11}(f)$

Consider the reduced system $S_5^{11}(f)$. We have

$$\left(S_5^{11}(f)\right) = \begin{cases} \alpha_1 = dy_1 + y_3 dy_4, \\ \alpha_2 = dy_2 - y_5 dy_3 + f dy_4, \\ \alpha_3 = dy_3 - \dfrac{\partial f}{\partial y_5} dy_4. \end{cases}$$

The differential forms α_4 and α_5 are arbitrary, we can assume that we have:

$$\begin{cases} \alpha_4 = dy_4, \\ \alpha_5 = dy_5. \end{cases}$$

The equation $d\alpha_2 = \alpha_3 \wedge \alpha_5 \mod (\alpha_1, \alpha_2)$ implies the following condition:

$$dy_3 \wedge dy_5 + df \wedge dy_4 = dy_3 \wedge dy_5 - \frac{\partial f}{\partial y_5} dy_4 \wedge dy_5 \quad \mathrm{mod}(\alpha_1, \alpha_2),$$

hence

$$y_5 \frac{\partial f}{\partial y_2} + \frac{\partial f}{\partial y_3} = 0, \tag{5.1}$$

and $d\alpha_3 = \alpha_4 \wedge \alpha_5 \mod (\alpha_1, \alpha_2, \alpha_3)$ implies

$$\frac{\partial^2 f}{\partial y_5^2} = 1.$$

Thus

$$\frac{\partial f}{\partial y_5} = y_5 + g(y_1, y_2, y_3, y_4)$$

and

$$f = \frac{1}{2}y_5^2 + y_5 g(y_1, y_2, y_3, y_4) + h(y_1, y_2, y_3, y_4).$$

According to 5.1 we can assume that

$$\frac{\partial g}{\partial y_2} = 0, \frac{\partial h}{\partial y_3} = 0 \quad \text{and} \quad \frac{\partial g}{\partial y_3} + \frac{\partial h}{\partial y_2} = 0.$$

Thus

$$f = \frac{1}{2}y_5^2 + y_5 g(y_1, y_3, y_4) + h(y_1, y_2, y_4).$$

A linear substitution of the form $\alpha_2' = \alpha_2 + \lambda\alpha_1$ with

$$\lambda = -\frac{g(y_1, y_3, y_4) - g(y_1, 0, y_4)}{y_3}$$

enables to assume that $\partial g/\partial y_3 = 0$, then $\partial h/\partial y_2 = 0$. We obtain

$$f = \frac{1}{2}y_5^2 + y_5 g(y_1, y_4) + h(y_1, y_4).$$

Theorem 5.5 *The Pfaffian system* $(S_5^{11}(f))$ *admits representatives of the form*

$$(S_5^{11}(f)) = \begin{cases} \alpha_1 = dy_1 + y_3 dy_4, \\ \alpha_2 = dy_2 - y_5 dy_3 + \left(\frac{1}{2}y_5^2 + y_5 g(y_1, y_4) + h(y_1, y_4)\right) dy_4, \\ \alpha_3 = dy_3 - (y_5 + g(y_1, y_4)) dy_4. \end{cases}$$

Let us determine the corresponding form $\mathcal{F}(\alpha_4, \alpha_5)$. We have

$$\begin{cases} d\alpha_1 = dy_3 \wedge dy_4, \\ d\alpha_2 = dy_3 \wedge dy_5 + (y_5 dy_5 + g dy_5 + y_5 \frac{\partial g}{\partial y_1} dy_1 + \frac{\partial h}{\partial y_1} dy_1) \wedge dy_4, \\ d\alpha_3 = -dy_5 \wedge dy_4 - \frac{\partial g}{\partial y_1} dy_1 \wedge dy_4, \end{cases}$$

then,

$$\begin{cases} d\alpha_1 = \alpha_3 \wedge \alpha_4 \\ d\alpha_2 = \alpha_3 \wedge \alpha_5 + \alpha_1 \wedge (y_5 g_1 + h_1)\alpha_4 \\ d\alpha_3 = \alpha_4 \wedge \alpha_5 - g_1\alpha_1 \wedge \alpha_4 \end{cases}$$

where $g_1 = \partial g/\partial y_1$ and $h_1 = \partial h/\partial y_1$ and more generally $g_{ijk} = \partial g/\partial y_i \partial y_j \partial y_k$. Let

$$\overline{\alpha}_5 = \alpha_5 + a\alpha_1,$$

where $a = a(y_1, y_4)$ is a function only of the variables y_1 and y_4. With respect to the basis $(\alpha_1, \alpha_2, \alpha_3, \alpha_4, \overline{\alpha}_5)$, we have

$$\begin{cases} d\alpha_1 = \alpha_3 \wedge \alpha_4, \\ d\alpha_2 = \alpha_3 \wedge \overline{\alpha}_5 + \alpha_1 \wedge [(y_5 g_1 + h_1)\alpha_4 + a\alpha_3], \\ d\alpha_3 = \alpha_4 \wedge \overline{\alpha}_5 + \alpha_1 \wedge [(-g_1 + a)\alpha_4], \\ d\alpha_4 = 0, \\ d\overline{\alpha}_5 = a\alpha_3 \wedge \alpha_4 + (a_1 y_3 - a_4)\alpha_1 \wedge \alpha_4. \end{cases}$$

By taking

$$a = \frac{2}{5}g_1$$

and

$$\begin{aligned} \pi_3 &= -\frac{2}{5}(a_1 y_3 - a_4)\alpha_1 + \frac{2}{5}g_1\alpha_3 + (g_1 y_5 + h_1)\alpha_4, \\ \pi_5 &= -\frac{3}{5}g_1\alpha_4, \\ \pi_i &= 0 \qquad i = 1, 2, 4, 6, 7, \end{aligned}$$

we obtain the Cartan' reduction

$$\begin{cases} d\alpha_1 = \alpha_3 \wedge \alpha_4, \\ d\alpha_2 = \alpha_3 \wedge \overline{\alpha}_5 + \alpha_1 \wedge \overline{\pi}_3, \\ d\alpha_3 = \alpha_4 \wedge \overline{\alpha}_5 + \alpha_1 \wedge \overline{\pi}_5, \\ d\alpha_4 = 0, \\ d\overline{\alpha}_5 = -\frac{4}{3}\alpha_3 \wedge \overline{\pi}_5 + \alpha_4 \wedge \overline{\pi}_3. \end{cases}$$

The calculations of the differential $d\pi_i$ enable us to determine the coefficients A_i, B_i. We have:

$$d\pi_1 = 0.$$

This implies

$$B_2 = B_3 = A_2 = A_3 = A_4 = 0.$$

Similarly, $d\pi_2 = 0$ is equivalent to

$$B_4 = A_4 = A_5 = 0.$$

But we have

$$
\begin{aligned}
d\pi_3 &= -\frac{3}{5}g_1\alpha_4 \wedge \bar{\alpha}_5 + (-\frac{16}{25}g_1^2 - \frac{2}{5}(y_3g_{111}-g_{141})y_3 + \frac{2}{5}g_{11}(g+y_5)) \\
&\quad +\frac{2}{5}((y_3g_{114}-g_{114}) + y_5g_{11} + h_{11})\alpha_1 \wedge \alpha_4 + (\frac{4}{5}g_{11})\alpha_1 \wedge \alpha_3.
\end{aligned}
$$

By comparing with the Cartan' equation

$$
\begin{aligned}
d\pi_3 &= \pi_4 \wedge (\pi_4 - \pi_1) - \alpha_5 \wedge \pi_5 + \alpha_2 \wedge \chi_1 \\
&\quad -B_1\alpha_1 \wedge \alpha_3 - A_1\alpha_1 \wedge \alpha_4 - A_2\alpha_1 \wedge \alpha_5,
\end{aligned}
$$

we obtain

$$
\begin{aligned}
A_1 &= \frac{16}{25}g_1^2 + \frac{2}{5}(y_3g_{111}-g_{141})y_3 - \frac{2}{5}g_{11}(g+y_5) \\
&\quad -\frac{2}{5}((y_3g_{114}-g_{114}) - y_5g_{11} - h_{11})
\end{aligned}
$$

and

$$
B_1 = -\frac{4}{5}g_{11}.
$$

Then the forms $\mathcal{F}$ and $\mathcal{G}$ are written

$\mathcal{F}(\alpha_4, \alpha_5) =$

$$
\left[\frac{16}{25}g_1^2 + \frac{2}{5}\left[(y_3g_{111}-g_{141})y_3 - g_{11}(g+y_5) - (y_3-1)g_{114} + y_5g_{11} + h_{11}\right]\right]\alpha_4^4
$$

$\mathcal{G}(\alpha_3, \alpha_4, \alpha_5) =$

$$
\left[\frac{16}{25}g_1^2 + \frac{2}{5}\left[(y_3g_{111}-g_{141})y_3 - g_{11}(g+y_5) - (y_3-1)g_{114} + y_5g_{11} + h_{11}\right]\right]\alpha_4^4
$$
$$
-\frac{16}{5}g_{11}\alpha_4^3\alpha_3.
$$

If $\mathcal{F}$ is identically non-zero then $\mathcal{F} = A_1\alpha_4^4$. The classification of Pfaffian systems $(S_{11}(f))$ summarizes at cases where A_1 is positive and $\mathcal{F}$ is a perfect square, and A_1 negative and $\mathcal{F}$ is, in absolute value, a perfect square.

Case where $\mathcal{F} = 0$

If $\mathcal{F}$ is identically zero then $\mathcal{G}$ is also zero. Thus

$$
g_{11} = 0
$$

hence

$$
g = y_1k(y_4) + l(y_4)
$$

and
$$h = -\frac{4}{5}y_1^2 k^2(y_4) + y_1 m(y_4) + n(y_4).$$
In the particular case where $g = 0$ and $h = 0$, the system (S) can be written
$$\left(S_5^{11}(f) \right) = \begin{cases} \alpha_1 = dy_1 + y_3 dy_4, \\ \alpha_2 = dy_2 - y_5 dy_3 + \frac{1}{2}y_5^2 dy_4, \\ \alpha_3 = dy_3 - y_5 dy_4. \end{cases}$$

5.5   Contact systems

5.5.1   Contact r-system

Integral manifolds

Let (S) be a Pfaffian system of constant rank r on $\mathbb{R}^n$. Let D be a completely integrable distribution tangent at each point to the distribution defined by (S). Let $(\omega_1, \cdots, \omega_r)$ be a basis of (S) and $(X_1, \cdots, X_q)$ vector fields linearly independent at each point and defining the distribution D. These vector fields satisfy
$$\begin{cases} [X_i, X_j] = \sum_{k=1}^q C_{ij}^k X_k, \\ \omega_i(X_j) = 0. \end{cases}$$
We know that if the system (S) is maximum class n, then the dimension q of the distribution D verifies $q \le r(n-r)/(r+1)$.

We are going to study the Pfaffian systems whose the kernel contains integrable distribution of maximum dimension $q = r(n-r)/(r+1)$.

Contact r-systems

Definition 5.1 *Let (S) be a Pfaffian system of rank r and of maximum class n on $\mathbb{R}^n$. We say that (S) is a contact r-system if there exists a completely integrable distribution D of dimension $q = r(n-r)/(r+1)$ contained in the kernel of (S).*

The existence of an integrable distribution of maximum dimension imposes obviously that $r(n-r)/(r+1)$ in an integer. By taking $rp = q$ we obtain
$$q(r+1) = r(n-r)$$
that is,
$$n = rp + p + r.$$

Example 21

For $n = 5$, $r = 2$, and $p = 1$, we consider the Pfaffian system

$$\begin{cases} \omega_1 = dx_1 + x_3 dx_4, \\ \omega_2 = dx_2 + x_5 dx_4. \end{cases}$$

The kernel of (S) is spanned by the vector fields

$$X_1 = \frac{\partial}{\partial x_3}, \quad X_2 = \frac{\partial}{\partial x_5}, \quad X_3 = \frac{\partial}{\partial x_4} - x_3 \frac{\partial}{\partial x_1} - x_5 \frac{\partial}{\partial x_2}.$$

Since $[X_1, X_2] = 0$ the distribution spanned by the vector fields X_1 and X_2 is completely integrable and contained in the kernel of this system. It is a contact 2-system.

Remark 10 *If $r = 1$ a contact 1-system is a Pfaffian system of rank 1 and of maximum class. By the Darboux theorem all contact 1-systems on $\mathbb{R}^n$ are locally isomorphic to the system defined by the equation*

$$\omega = dx_1 + x_2 dx_3 + \dots + x_{2p} dx_{2p+1} \quad (n = 2p + 1).$$

The main purpose of this paragraph is to generalize this theorem.

5.5.2 Classification of contact r-systems

Theorem 5.6 *Let (S) be a contact r−system on $\mathbb{R}^n$ with $n = r + p + rp$. There exists a local coordinate system*

$$(x_1, \dots, x_r, \ y_1, \dots, y_p, \ z_1, \dots, z_{pr})$$

of $\mathbb{R}^n$ so that the system (S) is represented by the following equations :

$$\begin{cases} \omega_1 = dx_1 + z_1 dy_1 + \dots + z_{(p-1)r+1} dy_p, \\ \omega_2 = dx_2 + z_2 dy_1 + \dots + z_{(p-1)r+2} dy_p, \\ \qquad \vdots \\ \omega_r = dx_r + z_r dy_1 + \dots + z_{pr} dy_p. \end{cases}$$

Proof. Let D be the distribution completely integrable contained in $\ker(S)$. It is possible to find a local coordinates system $(u_1, \dots, u_{rp}, \ v_1, \dots, v_{r+p})$ of $\mathbb{R}^n$ such that the vector fields $X_i = \partial/\partial u_i$

$(i = 1, ..., rp)$ generate the distribution D. Therefore we can choose as representatives of (S) the forms

$$
\begin{cases}
\omega_1 = dv_1 + \sum_{i=r+1}^{r+p} a_i^1 dv_i \\
\quad\quad\vdots \\
\omega_r = dv_r + \sum_{i=r+1}^{r+p} a_i^r dv_i.
\end{cases}
$$

Thus we have

$$
d\omega_j = \sum_{i=r+1}^{r+p} da_i^j \wedge dv_i,
$$

for every $j = 1, ..., r$. Since (S) is of maximum class n, the forms $\left(da_i^j, dv_i \right)$ $i = 1, ..., r + p$, $j = 1, ..., r$ are independent. It follows that the functions $\left(a_i^j, v_j \right)$ form a (local) coordinate system on $\mathbb{R}^n$. With respect to this coordinate system, the Pfaffian system admits the expected representation. This proves the theorem. ∎

5.6 $SO(n)$-Classification

5.6.1   Position of the problem

A Pfaffian system (S) of constant rank r on $\mathbb{R}^n$ is defined by a system of r linearly independent Pfaffian equations $\omega_1 = 0, ..., \omega_r = 0$. The kernel of this system is the distribution of $(n - r)$-planes defined by:

$$
\ker_x (S) = \{ X_x \in T_x R^n \,\backslash\, \omega_{ix} (X) = 0 \quad i = 1, ..., r \}.
$$

This distribution is the intrinsic object attached to (S). Note that the equations $\omega_i = 0$ do not form an intrinsic object attached to (S) but every system formed by independent equations defining the same distribution of $(n - r)$-planes. If by abuse of notation we were to denote $(S) = \{\omega_1, ..., \omega_r\}$ it would be necessary to understand in this notation that $\omega_1 = 0, ..., \omega_r = 0$ is a system of equations defining the distribution

$$
x \longmapsto \ker_x (S),
$$

defined on an open subset $U \subseteq \mathbb{R}^n$. Every other system of equations defining the Pfaffian system is of the form

$$\left(\alpha_i = \sum_{p=1}^{r} a_i^j \omega_j \right)_{1 \le i \le r}$$

where $M(x) = \left(a_i^j(x) \right)$ is an invertible matrix at each point. Naturally, this operation has been called a change of basis of (S), the matrices of the change of basis have values in $GL(n, \mathbb{R})$. Some differential geometry problems a privileged particular basis $(\omega_1, ..., \omega_r)$ of independent Pfaffian forms. In this case all changes of bases are not permitted if some properties attached to the coframe $(\omega_1, ..., \omega_r)$ have to be preserved. Consider the following typical example. Let g^2 be a Riemannian metric on $\mathbb{R}^n$. It can be written

$$g^2 = \omega_1^2 + ... + \omega_n^2$$

and $(\omega_1, ..., \omega_n)$ defines a Pfaffian system on $\mathbb{R}^n$ of rank n. Since this system is of rank n and of class n it is completely integrable, and by a change of basis we can assume that $\omega_i = dx_i$. The metric returns to the Euclidean metric

$$g^2 = dx_1^2 + ... + dx_n^2.$$

It is clear that this transformation returns the Riemannian geometry to Euclidean geometry alone . If we desire to preserve the properties of the metric g^2 we have to impose changes of bases preserving this metric, that is changes of bases with values in $SO(n)$ and not in $GL(n, \mathbb{R})$. Now the Pfaffian system associated to the metric cannot be reduced via the Frobenius theorem.

5.6.2 Classification of Pfaffian systems on $\mathbb{R}^2$ modulo $SO(2)$

We consider two Pfaffian forms ω_1 and ω_2 supposed to be linearly independent on $\mathbb{R}^2$. The associated equations $\omega_1 = 0$, $\omega_2 = 0$ define a Pfaffian system (S). The main goal of this section is to determine the local expression of (S) but the changes of bases allowed are of the form

$$\begin{cases} \tilde{\omega}_1 = a_1 \omega_1 + a_2 \omega_2, \\ \tilde{\omega}_2 = b_1 \omega_1 + b_2 \omega_2. \end{cases}$$

The matrix $M(x) = \begin{pmatrix} a_1(x) & a_2(x) \\ b_1(x) & b_2(x) \end{pmatrix}$ belongs to $SO(2)$ for every x.

Proposition 5.8 *By an $SO(2)$-change of basis the Pfaffian forms ω_1, ω_2 admit the following expression with respect to a local coordinates system (x,y) of $\mathbb{R}^2$*

$$\begin{cases} \omega_1 = \rho_1 dx, \\ \omega_2 = \mu dy + \rho_2 dx \end{cases}$$

where ρ_1 and μ are without zero functions.

Proof. With respect to the coordinate system (x,y) of $\mathbb{R}^2$, we have:

$$\begin{cases} \omega_1 = \alpha_1 dx + \alpha_2 dy, \\ \omega_2 = \beta_1 dx + \beta_2 dy. \end{cases}$$

Consider a matrix

$$M(x,y) = \begin{pmatrix} a_1(x,y) & a_2(x,y) \\ b_1(x,y) & b_2(x,y) \end{pmatrix} \in SO(2)$$

at each point, and take

$$\begin{cases} \widetilde{\omega}_1 = a_1\omega_1 + a_2\omega_2, \\ \widetilde{\omega}_2 = b_1\omega_1 + b_2\omega_2. \end{cases}$$

We have

$$\begin{cases} \widetilde{\omega}_1 = (a_1\alpha_1 + a_2\beta_1)\,dx + (a_1\alpha_2 + a_2\beta_2)\,dy, \\ \widetilde{\omega}_2 = (b_1\alpha_1 + b_2\beta_1)\,dx + (b_1\alpha_2 + b_2\beta_2)\,dy. \end{cases}$$

But $\omega_1 \wedge \omega_2 \neq 0$, the vector fields (α_2, β_2) and (α_1, β_1) are linearly independent at each point. Consider the unit vector field (a_1, a_2) normal to (α_2, β_2). For each unit vector field (b_1, b_2) normal to (a_1, a_2); the matrix

$$\begin{pmatrix} a_1 & a_2 \\ b_1 & b_2 \end{pmatrix}$$

belongs to $SO(2)$ and we have

$$\begin{aligned} \widetilde{\omega}_1 &= (a_1\alpha_1 + a_2\beta_1)\,dx \\ &= \rho_1 dx \end{aligned}$$

$$\begin{aligned} \widetilde{\omega}_2 &= (b_1\alpha_1 + b_2\beta_1)\,dx + (b_1\alpha_2 + b_2\beta_2)\,dy \\ &= \rho_2 dx + \mu dy. \end{aligned}$$

This proves the proposition. ∎

There exists an other possible reduction of the system:

Proposition 5.9 *There exists a local coordinate system* (u_1, u_2) *such that the forms* ω_1 *and* ω_2 *can be written:*

$$\begin{cases} \omega_1 = f(u_1, u_2)\, du_1, \\ \omega_2 = g(u_1, u_2)\, du_2. \end{cases}$$

Proof. In fact, we can find an integrating factor to ω_2 which is a function $K(x, y)$ such that

$$\mu(x, y)\, dy + \rho_2(x, y)\, dx = K(x, y)\, dv.$$

Since $\omega_1 \wedge \omega_2 \neq 0$ we have $dx \wedge dv \neq 0$ and with respect to the coordinate system (x, v) the Pfaffian forms ω_1 and ω_2 can be written in the forms desired . ■

Application: Study of Riemannian metric g^2 having same curvature tensor that the 'homogeneous' metric $\omega_1^2 + \omega_2^2$ and verifying

$$\begin{aligned} d\omega_1 &= \omega_1 \wedge \omega_2, \\ d\omega_2 &= 0. \end{aligned}$$

Let (α_1, α_2) the Pfaffian forms defined by

$$\alpha_1 = f\, dx,$$
$$\alpha_2 = g\, dy$$

where the functions f and g are without zero. We have

$$d\alpha_1 = \tfrac{\partial f}{\partial y} dy \wedge dx,$$
$$d\alpha_2 = \tfrac{\partial g}{\partial x} dx \wedge dy.$$

Thus

$$d\alpha_1 = \left(-\tfrac{1}{g}\tfrac{\partial f}{\partial y} dx + \tfrac{1}{f}\tfrac{\partial g}{\partial x} dy \right) \wedge \alpha_2,$$
$$d\alpha_2 = \left(\tfrac{1}{f}\tfrac{\partial g}{\partial x} dy + \tfrac{1}{g}\tfrac{\partial f}{\partial y} dx \right) \wedge \alpha_1.$$

Set

$$\beta_1 = -\frac{1}{g}\frac{\partial f}{\partial y}dx + \frac{1}{f}\frac{\partial g}{\partial x}dy,$$

then the curvature tensor is defined by:

$$\begin{aligned} d\beta_1 &= \alpha_1 \wedge \alpha_2 \\ &= \left(\frac{1}{g^2}\frac{\partial g}{\partial y}\frac{\partial f}{\partial y} - \frac{1}{g}\frac{\partial^2 f}{\partial y^2} + \frac{1}{f^2}\frac{\partial f}{\partial x}\frac{\partial g}{\partial x} - \frac{1}{f}\frac{\partial^2 g}{\partial x^2} \right) dy \wedge dx. \end{aligned}$$

We obtain

$$\frac{1}{g^2}\frac{\partial g}{\partial y}\frac{\partial f}{\partial y} - \frac{1}{g}\frac{\partial^2 f}{\partial y^2} + \frac{1}{f^2}\frac{\partial f}{\partial x}\frac{\partial g}{\partial x} - \frac{1}{f}\frac{\partial^2 g}{\partial x^2} = -fg.$$

We are going to determine the solutions of this equation by assuming $f = g$ (this case can always be assumed). Since f has no zero we can take $f = \exp(h)$. The previous equation is reduced to

$$(h_y)^2 - h_{yy} + (h_x)^2 - h_{xx} - (h_y)^2 - (h_x)^2 = e^{2h}$$

that is,

$$\triangle h = -e^{2h}.$$

Each solution of this equation defines a metric having the same curvature tensor as the Poincaré metric.

Chapter 6

k-SYMPLECTIC MANIFOLDS

In all of the following, smoothness should be understood to mean C^∞. The manifolds considered are Hausdorff and second countable.

6.1 Introduction

One of the main motivating factors in differential geometry is the conbined determination and classification of the differential manifolds equipped with a given geometrical structure.

Recently many works have beeb devoted to foliated manifolds, which we can see in the context of this book as smooth manifolds endowed with an integrable Pfaffian system. The study of contact and symplectic manifolds has received much attention. These manifolds are defined by the existence of a Pfaffian system of rank 1 and maximum class, or by a closed exterior 2-form of maximum rank.

It is possible to be interested in a more general framework concerning the study of manifolds equipped with a non-integrable Pfaffian system of a given constant class. For example the determination and the classification of connected (simply connected), possibly compact manifold of dimension 5 endowed with a structure locally defined by a Pfaffian system of the type $S_5^{11}(f)$, seems the most promising. Mathematical and physical considerations (the local study of Pfaffian systems and Nambu's statistical mechanics) have led us to an investigation of manifolds whose geometry is defined locally by k−symplectic systems. Recall also one of the results established in chapter 3 that the k−symplectic exterior systems are models of exterior systems of maximum rank. A general study of exterior systems needs of such k−symplectic systems. The purpose of this chapter is to define such ma-nifolds and to approach the existence theorems. The process is classic: we study the manifolds equipped with an almost k−symplectic structure (at every point of M the tangent space is equipped with a linear k-symplectic structure) and we make in obvious the conditions of integrability. The framework of the theory of G-structures is applied to the k-symplectic structure. A detailed exposition concerning the basic notions of differential geometry can be found in the book of Sternberg[52].

A multi-symplectic geometry in a classical field theory was initiated by Dedecker in 1953 and was developped by Tulczyjew around 1968, by Kijowski in 1973, Gawedzki in 1972 and Sczyba and Kondracki in 1979 (see for example [38] and [21]). The approach proposed by Geoffrey Martin [21] concerns a multi-symplectic model for electrodynamics based on the notion of symplectic $(k+1)$-vectors fields which generalizes the concept of cosymplectic structure on a Poisson manifold. He develops the necessary multilinear algebra of symplectic $(k+1)$-forms and uses these results to introduce the geometric structure on which the construction is based. He gives the dynamical condition and studies the formal properties which establish the relation between this construction and mechanic. Our step is analogous. But the basic notion is related to exterior system of forms of degree 2 when the approaches of Geoffrey Martin or Kijowski are based on an exterior form of degree greater than 3.

6.2 k-symplectic manifolds

6.2.1   Definition

Let M be a smooth manifold of dimension $n(k+1)$ equipped with a foliation $\mathfrak{F}$ of codimension n, and let $\theta^1, \ldots, \theta^k$ be closed differential forms on M of degree 2.

The sub-bundle of TM defined by the tangent vectors of leaves of the foliation $\mathfrak{F}$ will denoted by E, the set of all cross-sections of the M-bundle $TM \longrightarrow M$ (resp., $E \longrightarrow M$) will be denoted by $\mathfrak{X}(M)$ (resp., $\Gamma(E)$) and the set of all differential p-forms on M will be denoted by $\Lambda^p(M)$.

For every $x \in M$ we denote by $C_x(\theta^1), \ldots, C_x(\theta^k)$ the characteristic spaces of the 2-forms $\theta^1, \ldots, \theta^k$ at x. Recall that

$$C_x(\theta^p) = \{X_x \in T_xM \ \setminus \ i(X_x)\theta^p = 0 \text{ and } i(X_x)d\theta^p = 0\},$$

therefore,

$$C_x(\theta^p) = \{X_x \in T_x M \ \setminus \ i(X_x)\theta^p = 0\}$$

where $i(X_x)\theta^p$ denote the interior product of the vector X_x by the 2-form θ^p.

Definition 6.1 *We say that $\{\theta^1, \ldots, \theta^k\}$ is a k-symplectic differential system associated to the sub-bundle E (or that, the $(k+1)$-uple $(\theta^1, \ldots, \theta^k; E)$ is a k-symplectic structure on M), if the following conditions are satisfied*
1. The system $\{\theta^1, \ldots, \theta^k\}$ is non degenerate, that is,

$$C_x(\theta^1) \cap \cdots \cap C_x(\theta^k) = \{0\},$$

for every $x \in M$.
2. The system $\{\theta^1, \ldots, \theta^k\}$ is vanishing on the tangent vectors to the foliation $\mathfrak{F}$, that is,

$$\theta^p(X,Y) = 0$$

for all $X, Y \in \Gamma(E)$ and $p = 1, \ldots, k$.

6.2.2 Examples

1. Canonical k-symplectic structure on $\mathbb{R}^{n(k+1)}$

Consider the real space $\mathbb{R}^{n(k+1)}$ equipped with its Cartesian coordinates $(x^{pi}, x^i)_{1 \leq p \leq k, 1 \leq i \leq n}$. Let E be the sub-bundle of $T\mathbb{R}^{n(k+1)}$ defined by the equations

$$dx^1 = 0, \ \ldots \ , \ dx^n = 0$$

and let θ^p $(p = 1, \ldots, k)$ be the differential two forms on M given by

$$\theta^p = \sum_{i=1}^{n} dx^{pi} \wedge dx^i.$$

The $(k+1)$-tuple $(\theta^1, \ldots, \theta^k; E)$ defines a k-symplectic structure on $\mathbb{R}^{n(k+1)}$ called *the canonical k-symplectic structure*. This structure induces a natural k-symplectic structure on the torus $\mathbb{T}^{n(k+1)}$.

2. k-symplectic structure on the product of Lagrangian fibrations

Let

$$\xi^1 = (M^1, B, \pi^1), \ \ldots \ , \ \xi^k = (M^k, B, \pi^k)$$

be k smooth fibrations over an n–dimensional manifold B, let η be the trivial product

$$\eta = (M^1 \times \ldots \times M^k, B \times \ldots \times B, \pi^1 \times \ldots \times \pi^k)$$

and let δ be the diagonal mapping $x \longmapsto (x, \ldots, x)$ from B into $B \times \ldots \times B$. The fibre product

$$\xi = \xi^1 \times_B \ldots \times_B \xi^k$$

of fibrations $\xi^1, \ldots, \xi^k$ is the inverse image of η by δ. It is a smooth fibration $\xi = (M, B, \pi)$ whose the total space M is the set of elements $(x^1, \ldots, x^k) \in M^1 \times \ldots \times M^k$ such that $\pi^1(x^1) = \ldots = \pi^k(x^k)$; it is a closed submanifold of $M^1 \times \ldots \times M^k$. For every $b \in B$, the fibre $\pi^{-1}(b)$ is the product $(\pi^1)^{-1}(b) \times \ldots \times (\pi^k)^{-1}(b)$. Let

$$i : M \longrightarrow M^1 \times \ldots \times M^k$$

be the canonical inclusion mapping (i is an embedding) and

$$pr^p : M^1 \times \ldots \times M^k \longrightarrow M^p$$

$(p = 1, \ldots, k)$ the canonical projection on M^p. For every $p\,(p = 1, \ldots, k)$, the composite $pr^p \circ i$ is the restriction of pr^p to M; it is a submersion.

We suppose that ξ^p are Lagrangian fibrations (that is M^p is endowed with a symplectic structure σ^p such that the fibres define a Lagrangian foliation). Let $2n$ be the common dimension of manifolds M^p, then the fibres of ξ are of dimension nk and the manifold M is of dimension $n(k+1)$. For every $p\,(p = 1, \ldots, k)$, we take

$$\theta^p = (pr^p \circ i)^* \sigma^p.$$

Proposition 6.1 *In the previous hypothesis and notation the $(k+1)$-tuple $(\theta^1, \ldots, \theta^k; E)$ defines a k-symplectic structure on M, E is the sub-bundle of TM defined by the fibres of ξ.*

Proof. For every $p\,(p = 1, \ldots, k)$ we have

$$d\theta^p = (pr^p \circ i)^* d\sigma^p,$$

thus the forms θ^p are closed. Let X and Y be cross-sections of the sub-bundle E, let x be a point of M and let $p = 1, \ldots, k$. We have

$$\theta^p_x(X_x, Y_x) = \sigma^p_{(pr^p \circ i)(x)}((pr^p \circ i)^T_x X_x, (pr^p \circ i)^T_x Y_x),$$

where $(pr^p \circ i)^T$ is the tangent mapping associated with $(pr^p \circ i)$. The mappings $(pr^p \circ i)$ send the fibres of ξ onto the fibres of ξ^p, thus the vectors $(pr^p \circ i)_x^T X_x$ and $(pr^p \circ i)_x^T Y_x$ are vertical in M^p; the foliation defined by the fibres of ξ^p are Lagrangian by hypothesis, thus $\theta_x^p(X_x, Y_x) = 0$. If for each $x \in M$ and tangent vector X_x to M at x we have $i(X_x)\theta^p = 0$ for every p $(p = 1, \ldots, k)$, then

$$\theta_x^p(X_x, Y_x) = 0.$$

for all $p(p = 1, \ldots, k)$ and $Y_x \in T_x(M)$. The fact that for each $p\,(p = 1, \ldots, k)$, $(pr^p \circ i)$ is a submersion and σ^p is of rank $2n$ proves that $i_x^T X_x = 0$; thus $C_x(\theta^1) \cap \cdots \cap C_x(\theta^k) = \{0\}$. ∎

Notice that we have:

1. If the manifolds M^p are compact, it is similarly so for M and all leaves of the foliation defined by ξ are compact.

2. If $M^1, \ldots, M^k$ coincide with the cotangent bundle T^*B of a manifold B $(M^1 = \ldots = M^k = T^*B)$ we obtain a k-symplectic structure on the Whitney sum $T^*B \oplus \ldots \oplus T^*B$ over B.

3. k-symplectic structure on the bundle of k-(1-covelocities)

This example, often studied in the framework of the regular p-almost structure [41], corresponds to the canonical k-symplectic structure on the Whitney sum.

Let B be an n-dimensional manifold. We denote by T_{k1}^*B the cotangent bundle of k-(1-covelocities) of B, that is, the manifold of all 1-jets of mapping from B to $\mathbb{R}^k$ with target $0 \in \mathbb{R}^k$.

For each coordinate system $(x^j)_{1 \leq j \leq n}$ on B we associate the local coordinates

$$(x^i, x_i^1, \ldots, x_i^k)_{1 \leq i \leq n}$$

on T_{k1}^*B defined by

$$x^i(J_{x,0^1}f) = x^i(x),$$
$$x_i^p(J_{x,0^1}f) = (\frac{\partial f^p}{\partial x^i})(x),$$

where $J_{x,0^1}f$ is the 1-jet at $x \in B$ of the map

$$f = (f^1, \ldots, f^k) : B \longrightarrow \mathbb{R}^k$$

such that $f(x) = 0$.

We have $T^*_{k1} B$ is an $n(k+1)$-dimensional vector bundle with standard fibre type $\mathbb{R}^{nk}$; the canonical projection is the map $\pi : T^*_{k1} B \longrightarrow B$ defined by

$$\pi(J_{x,01} f) = x.$$

For each p $(p = 1, \ldots, k)$ we have a canonical 1-form λ^p defined by

$$\lambda^p(u)(X) = u^p(x)(\pi_* X)$$

where $u = (u^1, \ldots, u^k) \in T^*_{k1} B$, $X \in T_u(T^*_{k1} B)$ and $\pi(u) = x$. Let $\theta^p = d\lambda^p$ where $p = 1, \ldots, k$ and let $E = \ker \pi_*$; the $(k+1)$-tuple $(\theta^1, \ldots, \theta^k; E)$ defines a k-symplectic structure on $T^*_{k1} B$.

6.2.3 Darboux theorem

Theorem 6.1 *Let M be an $n(k+1)$-dimensional smooth manifold. If the $(k+1)$-tuple $(\theta^1, \ldots, \theta^k; E)$ is a k-symplectic structure on M then for every point p of M there exists an open neighborhood U of M containing p equipped with local coordinates $(x^{pi}, x^i)_{1 \leq p \leq k, 1 \leq i \leq n}$ called an adapted coordinate system, such that the differential forms θ^p are represented on U by*

$$\theta^p = \sum_{i=1}^{n} dx^{pi} \wedge dx^i,$$

and the sub-bundle E is defined by the equations

$$dx^1 = 0, \ldots, dx^n = 0.$$

Proof. It follows from the Frobenius theorem that there exists a system of local coordinates $(x_1, \ldots, x_{nk}, x^1, \ldots, x^n)$ defined on an open neighborhood U of M containing p such that the derivatives

$$\frac{\partial}{\partial x_1}, \ldots, \frac{\partial}{\partial x_{nk}}$$

generate the tangent space of the leaves at every point of U. The problem is of a local nature, therefore we can assume that U is an open neighborhood of $\mathbb{R}^{n(k+1)}$ and $p = (x_1(p), \ldots, x_{nk}(p), x^1(p), \ldots, x^n(p)) = 0$. The forms θ^p are locally exact (Poincaré's lemma), we can assume that the differential forms θ^p can be written on the open set U in the form

$$\theta^p = d \left(\sum_{u=1}^{nk} f^{pu} dx_u + \sum_{s=1}^{n} g^p_s dx^s \right)$$

where f^{pu} and g_s^p are smooth functions on U; thus

$$
\theta^p = \frac{1}{2} \sum_{1 \le u,v \le nk} \left(\frac{\partial f^{pu}}{\partial x_v} - \frac{\partial f^{pv}}{\partial x_u} \right) dx_u \wedge dx_v
$$
$$
+ \sum_{s=1}^{n} \left(\sum_{u=1}^{nk} \left(\frac{\partial g_s^p}{\partial x_u} - \frac{\partial f^{pu}}{\partial x^s} \right) dx_u + \frac{1}{2} \sum_{t=1}^{n} \left(\frac{\partial g_s^p}{\partial x^t} - \frac{\partial g_t^p}{\partial x^s} \right) dx^t \right) \wedge dx^s .
$$

The second condition of the definition of k-symplectic structure implies that

$$
\frac{\partial f^{pu}}{\partial x_v} = \frac{\partial f^{pv}}{\partial x_u}.
$$

For all $p = 1, \ldots, k$ and $i = 1, \ldots, n$ we take

$$
x^{pi} = g_i^p - \sum_{u=1}^{nk} \int_0^{x_j} \frac{\partial f^{pu}}{\partial x^i}(0, \ldots, 0, t, x_{j+1}, \ldots, x_{nk}, x^1, \ldots, x^n) dt.
$$

The relationship

$$
\theta^p = \sum_{i=1}^{n} dx^{pi} \wedge dx^i
$$

is equivalent to

$$
\frac{\partial x^{ps}}{\partial x_u} = \frac{\partial g_s^p}{\partial x_u} - \frac{\partial f^{pu}}{\partial x^s}
$$

and

$$
\frac{\partial x^{ps}}{\partial x^t} - \frac{\partial x^{pt}}{\partial x^s} = \frac{\partial g_s^p}{\partial x^t} - \frac{\partial g_t^p}{\partial x^s}.
$$

The above two relations are satisfied, thus

$$
\theta^p = \sum_{i=1}^{n} dx^{pi} \wedge dx^i.
$$

An argument analogous to that of the theorem of the classification of k−symplectic exterior systems proves that the Pfaffian forms dx^{pi} and dx^i are independent. ∎

Remark 11 *If* $(\theta^1, \ldots, \theta^k; E)$ *is a k-symplectic structure on M then the differential forms θ^p are of rank $2n$.*

6.2.4 Γ-structure associated with a k-symplectic system

Definition 6.2 *Let $\mathfrak{T}$ be a topological space. A pseudogroup Γ of homeomorphisms of $\mathfrak{T}$ is a family $(\varphi_i)_{i \in I}$ of homeomorphisms of $\mathfrak{T}$ with domain U_i and range V_i open subsets of $\mathfrak{T}$ such that:*

1. *the identity map of $\mathfrak{T}$ belongs to Γ;*

2. *if $\varphi_i \in \Gamma$ then its restriction to any open subset of its domain belongs to Γ;*

3. *if $\varphi_i \in \Gamma$, then $\varphi_i^{-1} \in \Gamma$;*

4. *if $\varphi_i : U_i \longrightarrow V_i$ and $\varphi_j : U_j \longrightarrow V_j$ belong to Γ, with $V_i \cap U_j \neq \emptyset$ then*

$$\varphi_j \circ \varphi_i : \varphi_i^{-1}(V_i \cap U_j) \longrightarrow \varphi_j(V_i \cap U_j)$$

   belongs to Γ;

5. *if $\varphi : U \longrightarrow V$ is local homeomorphism of $\mathfrak{T}$ which coincides on neighborhood of each point of U with an element of Γ then $\varphi \in \Gamma$.*

The pseudogroup Γ is called *transitive* if, in addition, for any two points of $\mathfrak{T}$, there is an element of Γ that takes the one to the other.

Definition 6.3 *Let Γ be a pseudogroup of homeomorphisms $\mathfrak{T}$. Let M be a topological space and let $\mathfrak{A}$ be a family $(\phi_i)_{i \in I}$ of local homeomorphisms*

$$\phi_i : U_i \longrightarrow V_i$$

of M onto $\mathfrak{T}$ where U_i (resp. V_i) is an open neighborhood of M (resp. $\mathfrak{T}$). We say that $\mathfrak{A}$ is a Γ-atlas, or an atlas compatible with Γ if the following conditions are satisfied:

1. $\bigcup_{i \in I} U_i = M$;

2. *if $\phi_i, \phi_j \in \mathfrak{A}$ then $\phi_i \circ \phi_j^{-1} \in \Gamma$.*

The elements of $\mathfrak{A}$ are called charts. A complete atlas of M compatible with Γ is an atlas compatible with Γ which is not contained in any other atlas of M compatible with Γ. Every atlas of M compatible with Γ is contained in a unique complete atlas compatible with Γ.

A complete Γ-atlas of M is called a Γ-structure.

Proposition 6.2 *Let M be an $n(k+1)$-dimensional smooth manifold. A k-symplectic structure on is equivalent to a given Γ-structure on M, where Γ is the pseudogroup of local diffeomorphisms of $\mathbb{R}^{n(k+1)}$ leaving the canonical k-symplectic structure of $\mathbb{R}^{n(k+1)}$ invariant.*

Proof. Suppose that M is equipped with a k-symplectic structure. It results from the Darboux theorem that there is an atlas $\mathfrak{A}$ (the Darboux atlas) whose the changes of coordinates belong to the pseudogroup Γ; in other words, the atlas $\mathfrak{A}$ defines a Γ-structure on M.

Conversely, if M admits a Γ-structure then there is an atlas $\mathfrak{A}$, whose changes of coordinates preserve the canonical k-symplectic structure $(\theta^1, \ldots, \theta^k; E)$ of $\mathbb{R}^{n(k+1)}$, that is, for all $(U, \varphi), (V, \psi) \in \mathfrak{A}$, we have:

1. $(\varphi \circ \psi^{-1})^* \theta^p = \theta^p$ for every $p (p = 1, \ldots, k)$,

2. $(\varphi \circ \psi^{-1})_* E(x) = E((\varphi \circ \psi^{-1})(x))$ for every $x \in \varphi(U) \cap \psi(V)$.

On the domain U of local chart (U, φ) of $\mathfrak{A}$ we take

$$\sigma(\varphi)^p = \varphi^* \theta^p$$

and

$$P(\varphi)(x) = (\varphi^{-1})_* E(\varphi(x)),$$

for all $p\ (p = 1, \ldots, k)$ and $x \in U$. On the intersection $U \cap V$ of the domains of the charts φ and ψ, we have

$$\sigma(\varphi)^p = \sigma(\psi)^p,$$
$$P(\varphi)(x) = P(\psi)(x).$$

for all $p\ (p = 1, \ldots, k)$ and $x \in U \cap V$; thus $(\sigma^1, \ldots, \sigma^k; P)$ defines a k-symplectic structure on M. ∎

6.3   Almost k-symplectic manifolds

6.3.1   Definition

Let M be a smooth manifold of dimension $n(k+1)$. We say that M is an almost k-symplectic manifold if for every $x \in M$ the tangent space $T_x M$ is equipped with a k-symplectic structure of vector spaces:

$$(\theta_x^1, \ldots, \theta_x^k; F_x),$$

where $\theta^p \in \Lambda^2(M)$.

Of course, we assume that this structure is smooth, that is, for every $x_0 \in M$ there exists an open neighborhood U_0 of x_0 in M and a smooth cross-section $(\omega^{pi}, \omega^i)_{1 \leq p \leq k, 1 \leq i \leq n}$ of the bundle of coframes of M such that

$$\theta^p = \sum_{i=1}^{n} \omega^{pi} \wedge \omega^i$$

and

$$F = \ker \omega^1 \cap \ldots \ker \omega^k$$

on U_0. Such a cross-section of the bundle of coframes of M will be called adapted to the almost k-symplectic structure of M.

6.3.2   Integrability of an almost k-symplectic structure

1. G-structure

Let G be a Lie group and let $\mathcal{G}$ be its Lie algebra. We assume that G acts on a smooth manifold M by a smooth right action $\phi : M \times G \longrightarrow M$. We have a linear mapping $X \longmapsto X^*$ of $\mathcal{G}$ into $\mathfrak{X}(M)$ where

$$X^*(x) = \left(\frac{d}{dt} \right)_{t=0} \phi(x, \exp(tX)),$$

for all $X \in \mathcal{G}$ and $x \in M$; we obtain

$$[X^*, Y^*] = [X, Y]^*$$

for all $X, Y \in \mathcal{G}$.

We often omit the ϕ and simply write $xg = R_g(x)$ for $\phi(x, g)$, for all $x \in M$ and $g \in G$.

Definition 6.4 *Let G be a Lie group with a unit element e, and let M be a smooth manifold. A principal G-bundle over M, with a total space P, with a base space M, and with a structural group G, is a smooth mapping $p : P \longrightarrow M$ together with a right action of G on P such that:*

1. $ug = u$ for some $g \in G$ implies that $g = e$;

2. M is covered by a family $(U_i)_{i \in I}$ of open neighborhoods such that for every $i \in I$ there is a diffeomorphism $\varphi_i : p^{-1}(U_i) \longrightarrow U_i \times G$ such that φ_i has the form $\varphi_i(u) = (p(u), \psi_i(u))$ and $\varphi_i(ug) = (p(u), \psi_i(u)g)$.

2. Bundle of Frames

Let M be a smooth manifold of dimension n. A frame at a point x of M is a linear isomorphism $u : \mathbb{R}^n \longrightarrow T_x M$. Let $\mathcal{F}_x$ be the set of all frames at x

$$\mathcal{F}_x = \{ u : \mathbb{R}^n \longrightarrow T_x M \setminus u \text{ is a linear isomorphism} \}$$

and let

$$\mathcal{F}_M = \bigcup_{x \in M} \mathcal{F}_x.$$

Let $p : \mathcal{F}_M \longrightarrow M$ be the mapping defined by $p(u) = x$ if u is a frame at x, that is $u \in \mathcal{F}_x$. The construction of the manifold structure on $\mathcal{F}_M$ is very similar to that already used for the tangent and cotangent bundles. Let (U, φ) be a chart on M. Define

$$\tilde{\varphi} : p^{-1}(U) \longrightarrow U \times \mathrm{GL}(n, \mathbb{R})$$

by

$$\tilde{\varphi}(u) = \left(p(u), d\varphi_{p(u)} \circ u \right).$$

$\tilde{\varphi}$ is a bijection and the topology and differential structure on $\mathcal{F}_M$ are defined by requiring these pairs $\left(p^{-1}(U), \tilde{\varphi} \right)$ to be charts. The group $\mathrm{GL}(n, \mathbb{R})$ acts on the right of $\mathcal{F}_M$ by $uA = u \circ A$ for $u \in \mathcal{F}_x$ and $A \in \mathrm{GL}(n, \mathbb{R})$. Note that

$$\tilde{\varphi}(uA) = \left(p(u), d\varphi_{p(u)} \circ u \circ A \right),$$

so the requirement of the previous definition are met. Thus, we have a principal $\mathrm{GL}(n, \mathbb{R})$-bundle.

On the bundle frames $\mathcal{F}_M$ there sits a canonically $\mathbb{R}^n$-valued linear differential form ω, defined by

$$\omega(X_u) = u^{-1}(p_* X_u)$$

for all $u \in \mathcal{F}_M$ and $X_u \in T_u(\mathcal{F}_M)$. The $\mathbb{R}^n$-valued linear differential form ω is called *the fundamental form* of the frames bundle $\mathcal{F}_M$.

Definition 6.5 *Let $p : P \longrightarrow M$ be a principal G-bundle and let H be a closed subgroup of G. A reduction of the structural group of P to H is a submanifold Q of P such that $p : Q \longrightarrow M$ is a principal H-bundle with restricted action (thus $QH = Q$).*

Definition 6.6 *Let* $p : P \longrightarrow M$ *be a principal G-bundle and let* $\mathcal{G}$ *be the Lie algebra of G. For* $g \in G$ *we denote by* R_g *the mapping from P into itself defined by* $R_g(u) = ug$. *By a connection on P we mean a* $\mathcal{G}$-*valued differential 1-form* π *such that:*

1. $\pi(u)(A^*(u)) = A$ *for all* $u \in P$ *and* $A \in \mathcal{G}$;

2. $\pi(ug) \circ R_g = Ad(g^{-1}) \circ \pi(u)$ *for all* $u \in P$ *and* $g \in G$.

With a linear connection π on the bundle frames we associate the torsion form of π, which is an $\mathbb{R}^n$-valued two form Ω defined on $\mathcal{F}_M$ and related with the fundamental form ω and the connection form π by

$$\Omega = d\omega + \pi \wedge \omega.$$

Definition 6.7 *Let* G *be a subgroup of* $GL(n, \mathbb{R})$. *A G-structure on an* $n-$*dimensional manifold M, is a G-reduction of* $\mathcal{F}(M)$ *to the group G. Thus a G-structure* B_G, *is a submanifold of* $\mathcal{F}(M)$ *with the property that for any* $x \in B_G$ *and any* $g \in GL(n, \mathbb{R})$ *the point xg belongs to* B_G *if and only if* $g \in G$.

A G-structure P on a smooth manifold M is said to be integrable if every point of M has a coordinate neighborhood U with local coordinate system $(x^1, \ldots, x^n)$ such that $(dx^1, \ldots, dx^n)$ is a cross-section of the bundle of the coframes $\mathcal{F}_M^*$ over U. We shall call such a local coordinate system $(x^1, \ldots, x^n)$ admissible with respect to the given G-structure P. If $(x^1, \ldots, x^n)$ and $(y^1, \ldots, y^n)$ are two admissible coordinate systems in open sets U and V respectively, then the Jacobian matrix $(\partial y^i / \partial x^j)_{1 \le i, j \le n}$ belongs to G at each point of $U \cap V$.

3. Adapted connection

Definition 6.8 *A linear connection* π *on an almost k-symplectic manifold is adapted to the almost k-symplectic structure if, with respect to an adapted cross-section of the bundle of coframes of M, the connection form* (π_v^u) *takes its values in the k-symplectic Lie algebra* $\mathfrak{sp}(k, n; \mathbb{R})$; *in other words, the components of the connection*

$$(\pi_j^s, \pi_j^{ps}, \pi_{pj}^s, \pi_{qj}^{ps})$$

with respect to an adapted cross-section satisfy

$$\begin{aligned}
\pi^s_{pj} &= 0, \\
\pi^{ps}_{qj} &= 0 \ \text{if} \ p \neq q, \\
\pi^{ps}_{pj} + \pi^j_s &= 0, \\
\pi^{ps}_j - \pi^{pj}_s &= 0.
\end{aligned}$$

The components Ω^u of the torsion Ω of the linear connection π are related to those of the connection form (π^u_f) and the fundamental form ω^u of the frames bundle by the relation

$$\Omega^u = d\omega^u + \sum_f \pi^u_f \wedge \omega^f$$

4. The case where $G = Sp(k, n; \mathbb{R})$

An almost k−symplectic structure on an $n(k+1)$−dimensional manifold M is equivalent to a given G-structure with $G = Sp(k, n; \mathbb{R})$. Such a $Sp(k, n; \mathbb{R})$-structure is integrable if this almost k−symplectic structure corresponds to a k−symplectic structure. Consequently we can return to the calculation of the Bernard tensor in order to integrate this G-structure. But the vanishing of this tensor is equivalent to the existence of an adapted connection without torsion. Therefore we are going to study the problem of the existence of such a connection.

Note that the integrability of a G-structure implies the vanishing of the Bernard tensor; the reverse is false in the general case. In the case of an almost k−symplectic structure there is an equivalence between the integrabi-lity and the vanishing of this tensor (or the existence of an adapted connection without torsion).

The almost k-symplectic structure is integrable if and only if about every point of M we can find a coordinate neighborhood U with a local coordinate system $(x^{pi}, x^i)_{1 \leq p \leq k, 1 \leq i \leq n}$ such that

$$\theta^p = \sum_{i=1}^n dx^{pi} \wedge dx^i$$

and F is defined by the equations

$$dx^1 = 0, \ldots, \ dx^n = 0$$

at each point of U.

Proposition 6.3 *Let M be an $n(k+1)$-dimensional manifold equipped with an almost k-symplectic structure such that the distribution $x \longmapsto F(x)$ is integrable. Then the almost k-symplectic structure is integrable if and only if the manifold M admits an adapted connection without torsion.*

Proof. Let π be an adapted connection without torsion, then for any adapted cross-section $(\omega^{pi}, \omega^i)_{1 \leq p \leq k, 1 \leq i \leq n}$ we have

$$d\omega^{pj} = -\sum_{s=1}^{n} (\pi_s^{pj} \wedge \omega^s + \pi_{ps}^{pj} \wedge \omega^{ps})$$

and

$$d\omega^j = -\sum_{s=1}^{n} \pi_s^j \wedge \omega^s.$$

The differential forms θ^p vanish; in fact, we have

$$d\theta^p = -\sum_{s,j=1}^{n} (\pi_s^{pj} \wedge \omega^s \wedge \omega^j + \pi_{ps}^{pj} \wedge \omega^{ps} \wedge \omega^j - \omega^{pj} \wedge \pi_s^j \wedge \omega^s).$$

It results from the Darboux theorem that every point of M has a coordinate neighborhood U with coordinate system $(x^{pi}, x^i)_{1 \leq p \leq k, 1 \leq i \leq n}$, such that:

$$\theta^p = \sum_{i=1}^{n} dx^{pi} \wedge dx^i$$

and the sub-bundle F is defined by the equations $dx^1 = 0, \ldots, dx^n = 0$. ∎

Remark 12 *An integrable $Sp(k, n; \mathbb{R})$-structure is of infinite type because the Lie algebra $\mathfrak{sp}(k, n; \mathbb{R})$ contains a matrix of rank 1.*

6.4　Hamiltonian systems

In this section we extend the classic formalism of symplectic geometry to the k-symplectic structure.

6.4.1　Basic functions-foliated vector fields

Let M be an n-dimensional smooth manifold, let $\mathfrak{F}$ be a q-codimensional foliation on M, and let E be its corresponding p-dimensional integrable sub-bundle of TM, with $p = n - q$.

Definition 6.9 *A smooth function f on M is said to be basic if for every $Y \in \Gamma(E)$ the derivative $Y(f)$ of f along Y is identically zero. We will denote the set of the basic functions by $\Lambda_b^0(M, \mathfrak{F})$;it is clearly a subring of the ring $\Lambda^0(M)$ of smooth functions.*

Proposition 6.4 *Let $f \in \Lambda^0(M)$. The following properties are equivalent*
  1. f is basic;
  2. f is constant on each leaf;
  3. In every simple distinguished open set equipped with distinguished local coordinates $(x^1, ..., x^p, y^1, ..., y^q)$ the function f depends only on the variables $y^1, ..., y^q$.

If U is an arbitrary open set of M, a smooth function f on U which is basic for $\mathfrak{F}_U$ will be said to be a basic local function.

Definition 6.10 *A vector field on M is said to be foliated, or an infinitesimal automorphism of $\mathfrak{F}$, if for all $Y \in \Gamma(E)$ the Lie bracket $[X, Y]$ also belongs to $\Gamma(E)$.*

Proposition 6.5 *Let $X \in \mathfrak{X}(M)$. The following properties are equivalent*
  1. X is foliated;
  2. If $(\varphi_t)_{|t|<\varepsilon}$ is the local one parameter group associated with X on neighborhood of an arbitrary point of M, then, for all t, the local diffeomorphism φ_t leaves the sub-bundle E invariant;
  3. In every simple distinguished open set equipped with distinguished local coordinates $(x^i, y^j)_{1 \le i \le p, \ 1 \le j \le q}$,the vector field X takes the form:

$$X = \sum_{s=1}^{p} \xi^s(x^1, \ldots, x^p, y^1, \ldots, y^q) \frac{\partial}{\partial x^s} + \sum_{t=1}^{q} \eta^t(y^1, \ldots, y^q) \frac{\partial}{\partial y^t}.$$

We denote by $\mathfrak{L}(M, \mathfrak{F})$ the Lie algebra of foliated vector fields for $\mathfrak{F}$. Note that if $f \in \Lambda_b^0(M, \mathfrak{F})$ and $X \in \mathfrak{L}(M, \mathfrak{F})$ the function $X(f)$ is basic ($X(f) \in \Lambda_b^0(M, \mathfrak{F})$) and the vector field fX is foliated ($fX \in \mathfrak{L}(M, \mathfrak{F})$).

6.4.2 Hamiltonian systems

Let M be a smooth manifold of dimension $n(k + 1)$ equipped with a k-symplectic structure $(\theta^1, \ldots, \theta^k; E)$ and let j be the mapping from $\mathfrak{X}(M)$ into $\Lambda^1(M) \times \ldots \times \Lambda^1(M)$ defined by

$$j(X) = (j^1(X), \ldots, j^k(X)) = (i(X)\theta^1, \ldots, i(X)\theta^k)$$

for every $X \in \mathfrak{X}(M)$.

Definition 6.11 *A vector field* X *on* M *is called a Hamiltonian system if it is an infinitesimal automorphism for the k-symplectic structure* $(\theta^1, \ldots, \theta^k; E)$, *that is, if the following conditions are satisfied:*

1. X *is foliated;*

2. $L_X\theta^1 = \ldots = L_X\theta^k = 0$ *where* L_X *is the Lie derivative with respect to* X.

From the relationship

$$L_X\theta = i(X)d\theta + di(X)\theta$$

we see that the second condition of the previous definition is equivalent to requiring that the Pfaffian forms $i(X)\theta^1, \ldots, i(X)\theta^k$ are closed.

6.4.3   Hamiltonian mappings

Let X be a Hamiltonian system. It results from the Poincaré lemma that for every $x \in M$ there exists an open neighborhood U of M containing x and a smooth mapping $H : U \longrightarrow \mathbb{R}^k$ satisfying the relation

$$j(X)_{|U} = -dH_{|U}.$$

Conversely, if H is a smooth mapping from M into $\mathbb{R}^k$ such that $dH \in j(\mathcal{L}(M, \mathfrak{F}))$, there exists a unique vector field on M, denoted X_H and called the Hamiltonian system associated with H, such that

$$j(X_H) = -dH$$

on M; the vector field X_H is called *a strongly Hamiltonian system.*

A smooth mapping $H : M \longrightarrow \mathbb{R}^k$ satisfying $dH \in j(\mathcal{L}(M, \mathfrak{F}))$ is called a Hamiltonian mapping of the k-symplectic structure $(\theta^1, \ldots, \theta^k; E)$.

Proposition 6.6 *Let* $H = (H^p)_{1 \leq p \leq k}$ *be a Hamiltonian mapping and let* X_H *be the associated Hamiltonian system. With respect to an adapted coordinates system* $(x^{pi}, x^i)_{1 \leq p \leq k, 1 \leq i \leq n}$, *the components* H^p *of* H *and* X_H *can be written:*

$$H^p = \sum_{j=1}^{n} f_j(x^1, \ldots, x^n)x^{pj} + g(x^1, \ldots, x^n)$$

and

$$X_H = -\sum_{s=1}^{n}\sum_{p=1}^{k}\left(\sum_{j=1}^{n}\frac{\partial f_j}{\partial x^s}(x^1, \ldots, x^n)x^{ps} + \frac{\partial g^p}{\partial x^s}(x^1, \ldots, x^n)\right)\frac{\partial}{\partial x^{ps}}$$

$$+ \sum_{s=1}^{n} f_s(x^1, \ldots, x^n)\frac{\partial}{\partial x^s},$$

where f_j and g^p are smooth basic functions on U.

Proof. Let U be an open neighborhood of M equipped with an adapted coordinate system $(x^{pi}, x^i)_{1 \leq p \leq k, 1 \leq i \leq n}$. Let

$$X_H = \sum_{s=1}^{n} \sum_{p=1}^{k} X_{ps} \frac{\partial}{\partial x^{ps}} + \sum_{j=1}^{n} X_j \frac{\partial}{\partial x^j}$$

be the expression of the Hamiltonian system X_H in these coordinates. The relationships

$$j^r(X_H) = i(X_H)\theta^r = \sum_{j=1}^{n}(X_{rj}dx^j - X_j dx^{rj}) = -dH^r \quad (r = 1, \ldots, k)$$

imply that

$$X_{rj} = -\frac{\partial H^r}{\partial x^j}, \quad X_s = \frac{\partial H^r}{\partial x^{rs}}, \quad \frac{\partial H^r}{\partial x^{ps}} = 0 \text{ if } p \neq r.$$

But X is a foliated vector field, thus we have:

$$\frac{\partial X_s}{\partial x^{ps}} = 0,$$

and the components H^p of H take the form

$$H^p = \sum_{j=1}^{n} f_j(x^1, \ldots, x^n)x^{pj} + g^p(x^1, \ldots, x^n)$$

where f_j and g^p are smooth functions on U depending only on the variables $x^1, \ldots, x^n$. ∎

Remark 13 *Assume that $k \geq 2$. It follows from the proof of the previous proposition that, if the Pfaffian forms $i(X)\theta^1, \ldots, i(X)\theta^k$ are closed (or equivalently $L_X\theta^1 = \ldots = L_X\theta^k = 0$) the vector field X is necessarily an infinitesimal automorphism of $\mathfrak{F}$.*

6.4.4   Poisson bracket

Proposition 6.7 *Let H and K be two Hamiltonian mappings and X_H, X_K the associated Hamiltonian systems. The Lie bracket $[X_H, Y_K]$ is a Hamiltonian system. More precisely, the mapping denoted $\{H, K\}$ of M into $\mathbb{R}^k$ defined by*

$$\{H, K\} = -(\theta^1(X_H, X_K), \ldots, \theta^k(X_H, X_K))$$

satisfies

$$[X_H, X_K] = X_{\{H,K\}}.$$

The mapping $\{H, K\}$ is called the Poisson bracket of the Hamiltonian mappings H and K.

Proof. In fact, we have

$$i([X_H, X_K])\theta^p = [\mathcal{L}_{X_H}, i(X_K)]\theta^p = d(\theta^p(X_H, X_K)) = -d\{H, K\}^p$$

for every $p(p = 1, \ldots, k)$. With respect to an adapted coordinate system $(x^{pi}, x^i)_{1 \le p \le k, 1 \le i \le n}$ the components $\{H, K\}^p$ of $\{H, K\}$ are given by:

$$\{H, K\}^p = \sum_{s=1}^{n} \left(\frac{\partial H^p}{\partial x^s} \frac{\partial K^p}{\partial x^{ps}} - \frac{\partial H^p}{\partial x^{ps}} \frac{\partial K^p}{\partial x^s} \right),$$

which completes the proof. ∎

Let $\mathfrak{H}(M)$ be the set of Hamiltonian mappings of the k-symplectic structure $(\theta^1, \ldots, \theta^k; E)$. The association $(H, K) \longmapsto \{H, K\}$, of $\mathfrak{H}(M) \times \mathfrak{H}(M)$ into $\mathfrak{H}(M)$, is a skew-symmetric $\mathbb{R}$-bilinear mapping satisfying the Jacobi identity.

Proposition 6.8 $(\mathfrak{H}(M), \{,\})$ *is an infinite-dimensional Lie algebra.*

6.5   Nambu's statistical mechanics

A notable feature of the Hamiltonian description of classical dynamics is Liouville's theorem, which states that the volume of phase space occupied by an ensemble of systems is conserved. The theorem plays, amongst other things, a fundamental role in statistical mechanics. On the other hand, Hamiltonian dynamics is not the only formalism that makes a statistical mechanics possible. Any set of equations which lead to a Liouville theorem in a suitably defined phase space will do (provided, of course, that ergodicity may be assumed). Nambu proposes a possible generalization of the Hamiltonian dynamics for a 3-dimensional space.

In this section we are going to see that the Hamiltonian mappings of the 2-symplectic structure of $\mathbb{R}^3$ are solutions of Nambu's system. First, we recall the basic equations of Nambu's statistical mechanics [47].

Let (x, y, z) be a triplet of dynamical variables (a canonical triplet) which span a 3-dimensional phase space M. This is a formal generalization of conventional phase space spanned by a canonical pair (p, q). Next, we will introduce two functions H and G depending on (x, y, z) which serve as a pair of Hamiltonians to determine the motion of points in phase space. More precisely Nambu has postulated the following Hamilton equation:

$$\frac{dx}{dt} = \frac{D(H,G)}{D(y,z)},$$

$$\frac{dy}{dt} = \frac{D(H,G)}{D(z,x)},$$

$$\frac{dz}{dt} = \frac{D(H,G)}{D(x,y)},$$

where $D(H,G)/D(y,z)$ denote the Jacobian

$$\frac{D(H,G)}{D(y,z)} = \frac{\partial H}{\partial y}\frac{\partial G}{\partial z} - \frac{\partial H}{\partial z}\frac{\partial G}{\partial y}.$$

The above equations are called *Nambu's equations of motion*, and the vector field whose integral curves are given by Nambu's equations of motion will be denoted by $X^n_{(H,G)}$ and called the *dynamical system of Nambu*.

Consider the space $M = \mathbb{R}^3$ equipped with its canonical 2-symplectic structure $(\theta^1, \theta^2; E)$ defined by

$$\theta^1 = dx \wedge dz,$$
$$\theta^2 = dy \wedge dz,$$
$$E = \ker dz.$$

The Hamiltonian mapping of the 2-symplectic structure are the mapping

$$H : M \longrightarrow \mathbb{R}^2$$

whose components are given by

$$H^1 = f(z)x + g^1(z),$$
$$H^2 = f(z)y + g^2(z),$$

where f, g^1 and g^2 are smooth real functions depending only on the variable z. The integral curves of the Hamiltonian system X_H of the 2-symplectic structure are given by the following equations:

$$\frac{dx}{dt} = -\frac{\partial H^1}{\partial z},$$

$$\frac{dy}{dt} = -\frac{\partial H^2}{\partial z},$$

and

$$\frac{dz}{dt} = \frac{\partial H^1}{\partial x} = \frac{\partial H^2}{\partial y}.$$

Theorem 6.2 Let $H = (H^1, H^2)$ with $H^1 = f(z)x + g^1(z)$ and $H^2 = f(z)y + g^2(z))$ be Hamiltonian mappings of the 2-symplectic structure. Then the Hamiltonian system X_H and the dynamical system of Nambu X_H^n are related by

$$X_H^n = f(z)X_H.$$

Proof. This is an immediate consequence of the following relations:

$$\frac{D(H^1, H^2)}{D(y, z)} = -f(z)\frac{\partial H^1}{\partial z} \; , \quad \frac{D(H^1, H^2)}{D(z, x)} = -f(z)\frac{\partial H^2}{\partial z} \; ,$$

and

$$\frac{D(H^1, H^2)}{D(x, y)} = -f(z)\frac{\partial H^1}{\partial x} = -f(z)\frac{\partial H^2}{\partial y}.$$

This completes the proof of this proposition. ■

Corollary 6.1 The mapping

$$(f(z))^{-1}H = (x + h^1(z), y + h^2(z))$$

is a solution of Nambu's equations of motions on a domain where $f(z)$ is a non-vanishing function and

$$h^1(z) = (f(z))^{-1}g^1(z) \quad , \quad h^2(z) = (f(z))^{-1}g^2(z).$$

6.6   k-symplectic Lie algebras

6.6.1   Definition

Let G be a Lie group of dimension $n(k+1)$ with Lie algebra $\mathcal{G}$, equipped with a k-symplectic structure $(\theta^1, \ldots, \theta^k; E)$. We assume that such a structure is left-invariant that is, for any $g, x \in G$ and $p = 1, \ldots, k$ we have

1. $L_g^* \theta^p = \theta^p$,

2. $(L_g)_* E(x) = E(gx)$.

Let $\mathfrak{F}$ be the left-invariant foliation defined by the sub-bundle E. The leaf H of $\mathfrak{F}$ passing through the unit element e of G is an n- codimensional connected Lie subgroup of G. Let $\mathcal{H}$ be the Lie algebra of the subgroup H. Thus, we have associated with the sub-bundle E, a Lie subalgebra $\mathcal{H}$ of $\mathcal{G}$ such that

$$\theta^p(X, Y) = 0$$

for all $X, Y \in \mathcal{H}$; we identify the differential form $\theta^p(e)$ at the unit element of G with the corresponding bilinear form on $\mathcal{G}$, also denoted θ^p. Conversely, let $\mathcal{H}$ be a subalgebra of $\mathcal{G}$. For every $x \in G$ we take:

$$P(x) = \{X_x \in T_a G \mid \tilde{\eta}(X_x) \in \mathcal{H}\}$$

where $\tilde{\eta}$ is the Maurer–Cartan 1-form of G. The associated distribution defines an integrable sub-bundle of TG. The leaf H of the foliation $\mathfrak{F}$ associated with P and passing through the unit element e of G, is a Lie subgroup H of G with Lie algebra $\mathcal{H}$ such that the orbits of H acting by the right translations on G are the leaves of $\mathfrak{F}$.

Definition 6.12 *Let $\mathcal{G}$ be a Lie algebra of dimension $n(k+1)$ over $\mathbb{K}$ ($\mathbb{K} = \mathbb{R}$ or $\mathbb{C}$), $\theta^1, \ldots, \theta^k$ be closed forms of degree two (belonging to $\Lambda^2(\mathcal{G})$), and let $\mathcal{H}$ be a Lie subalgebra of $\mathcal{G}$ of codimension n. We say that $(\theta^1, \ldots, \theta^k; \mathcal{H})$ is a k-symplectic structure on $\mathcal{G}$ if the following conditions are satisfied*
  1. The exterior system $\{\theta^1, \theta^2, \cdots, \theta^k\}$ is non-degenerate, that is

$$A(\theta^1) \cap \cdots \cap A(\theta^k) = (0);$$

  2. The Lie subalgebra $\mathcal{H}$ is a totally isotropic subspace of $\mathcal{G}$ with respect the system $\{\theta^1, \theta^2, \cdots, \theta^k\}$, that is,

$$\theta^p(x, y) = 0$$

for all x, y belonging to $\mathcal{H}$ and $p = 1, \ldots, k$.

Let $\mathcal{A}$ be an $n(k+1)$-dimensional vector space over the field $\mathbb{K}$ ($\mathbb{K} = \mathbb{R}$ or $\mathbb{C}$) and let $(\omega^j)_{1 \leq j \leq n(k+1)}$ be a basis of the dual space $\mathcal{A}^*$ of $\mathcal{A}$. Considering the mapping δ of $\mathcal{A}^*$ into $\Lambda^2(\mathcal{A}^*)$ defined by:

$$\delta \omega^r = \sum_{i=1}^{n-1} \omega^{ik+r} \wedge \omega^{nk+i+1} \quad \text{for } 1 \leq i \leq k,$$
$$\delta \omega^a = 0 \quad \text{if } k+1 \leq a \leq nk+1,$$
$$\delta \omega^{nk+i} = -\omega^{nk+1} \wedge \omega^{nk+i} \quad \text{for } 2 \leq i \leq n.$$

We have $\delta \circ \delta = 0$, therefore the mapping δ confers upon $\mathcal{A}$ a Lie algebras law. Let $(X_j)_{1 \leq j \leq n(k+1)}$ be the basis of $\mathcal{A}$ whose dual basis is $(\omega^j)_{1 \leq j \leq n(k+1)}$, and let $\mathcal{J}$ be the ideal of $\mathcal{A}$ spanned by $(X_j)_{1 \leq j \leq nk}$. The $(k+1)$-tuple

$$(\theta^1 = \delta\omega^1, \ldots, \theta^k = \delta\omega^k; \mathcal{J})$$

defines a k-symplectic structure on $\mathcal{A}$.

6.6.2   Nilpotent 1-symplectic Lie algebras

Recall here the theorem of Medina ([43]).

Theorem 6.3 *Let G be a Lie group with a completely solvable Lie algebra $\mathcal{G}$. If Ω is a left-invariant symplectic form on G, then (G, Ω) admits an invariant Lagrangian foliation.*

Recall that a Lie algebra $\mathcal{G}$ is completely solvable if the adjoint representation of $\mathcal{G}$ is trigonalizable, in other hand, if there exists a decreasing sequence of ideals (of $\mathcal{G}$) of dimensions n, $n-1$, $n-2, \ldots$, 0, where $n = \dim \mathcal{G}$. For example every nilpotent Lie algebra is completely solvable.

Corollary 6.2 *Every symplectic Lie algebra is 1-symplectic.*

Consequence: Classification of 1-symplectic nilpotent Lie algebra of dimension less than 6.

1. $\dim \mathcal{G} = 2$.

The Lie algebra is abelian. The equations of structure of $\mathcal{G}$ are given by:

$$d\alpha_i = 0, \ (i = 1, 2).$$

2. $\dim \mathcal{G} = 4$.

2.1 The abelien Lie algebra whose equations of structure are given by

$$d\alpha_i = 0, \quad (i = 1, \ldots, 4).$$

2.2 The Lie algebra $h_3 \oplus a$ defined by:

$$d\alpha_1 = \alpha_2 \wedge \alpha_3$$
$$d\alpha_i = 0, \quad (i = 2, 3, 4).$$

2.3 The Lie algebra n_4 defined by:

$$d\alpha_3 = \alpha_1 \wedge \alpha_2$$
$$d\alpha_4 = \alpha_1 \wedge \alpha_3$$
$$d\alpha_i = 0., \quad i = 2, 3.$$

3. $\dim \mathcal{G} = 6$.

Every nilpotent 1-symplectic Lie algebra is isomorphic to one of the following Lie algebras:

$$
n_{6,3} = \begin{cases}
d\alpha_3 = \alpha_1 \wedge \alpha_2, \\
d\alpha_4 = \alpha_1 \wedge \alpha_3, \\
d\alpha_5 = \alpha_1 \wedge \alpha_4 + \alpha_2 \wedge \alpha_3, \\
d\alpha_6 = \alpha_1 \wedge \alpha_5 + \alpha_2 \wedge \alpha_4;
\end{cases}
$$

$$
n_{6,4} = \begin{cases}
d\alpha_3 = \alpha_1 \wedge \alpha_2, \\
d\alpha_4 = \alpha_1 \wedge \alpha_3, \\
d\alpha_5 = \alpha_1 \wedge \alpha_4, \\
d\alpha_6 = \alpha_1 \wedge \alpha_5 + \alpha_2 \wedge \alpha_3;
\end{cases}
$$

$$
n_{6,5} = \begin{cases}
d\alpha_3 = \alpha_1 \wedge \alpha_2, \\
d\alpha_4 = \alpha_1 \wedge \alpha_3, \\
d\alpha_5 = \alpha_1 \wedge \alpha_4, \\
d\alpha_6 = \alpha_1 \wedge \alpha_5;
\end{cases}
$$

$$
n_{6,6} = \begin{cases}
d\alpha_3 = \alpha_1 \wedge \alpha_2, \\
d\alpha_4 = \alpha_1 \wedge \alpha_3, \\
d\alpha_5 = \alpha_2 \wedge \alpha_3, \\
d\alpha_6 = \alpha_1 \wedge \alpha_4 + \alpha_2 \wedge \alpha_5;
\end{cases}
$$

$$
n_{6,7} = \begin{cases}
d\alpha_3 = \alpha_1 \wedge \alpha_2, \\
d\alpha_4 = \alpha_1 \wedge \alpha_3, \\
d\alpha_5 = \alpha_2 \wedge \alpha_3, \\
d\alpha_6 = \alpha_1 \wedge \alpha_4 - \alpha_2 \wedge \alpha_5;
\end{cases}
$$

$$
n_{6,8} = \begin{cases}
d\alpha_3 = \alpha_1 \wedge \alpha_2, \\
d\alpha_4 = \alpha_1 \wedge \alpha_3, \\
d\alpha_5 = \alpha_1 \wedge \alpha_4, \\
d\alpha_6 = \alpha_2 \wedge \alpha_3;
\end{cases}
$$

$$
n_{6,9} = \begin{cases}
d\alpha_4 = \alpha_1 \wedge \alpha_2, \\
d\alpha_5 = \alpha_1 \wedge \alpha_3 + \alpha_2 \wedge \alpha_4, \\
d\alpha_6 = \alpha_2 \wedge \alpha_5 + \alpha_3 \wedge \alpha_4,
\end{cases}
$$

$$n_{6,10} = \begin{cases} d\alpha_4 = \alpha_1 \wedge \alpha_2, \\ d\alpha_5 = \alpha_1 \wedge \alpha_4, \\ d\alpha_6 = \alpha_1 \wedge \alpha_5 + \alpha_2 \wedge \alpha_3 + \alpha_2 \wedge \alpha_4, \end{cases}$$

$$n_{6,11} = \begin{cases} d\alpha_4 = \alpha_1 \wedge \alpha_2, \\ d\alpha_5 = \alpha_1 \wedge \alpha_4, \\ d\alpha_6 = \alpha_1 \wedge \alpha_5 + \alpha_2 \wedge \alpha_3; \end{cases}$$

$$n_{6,12} = \begin{cases} d\alpha_4 = \alpha_1 \wedge \alpha_2, \\ d\alpha_5 = \alpha_1 \wedge \alpha_3 + \alpha_1 \wedge \alpha_4, \\ d\alpha_6 = \alpha_2 \wedge \alpha_4; \end{cases}$$

$$n_{6,13} = \begin{cases} d\alpha_4 = \alpha_1 \wedge \alpha_2, \\ d\alpha_5 = \alpha_1 \wedge \alpha_3 + \alpha_2 \wedge \alpha_4, \\ d\alpha_6 = \alpha_1 \wedge \alpha_4; \end{cases}$$

$$n_{6,14} = \begin{cases} d\alpha_4 = \alpha_1 \wedge \alpha_2, \\ d\alpha_5 = \alpha_1 \wedge \alpha_3 - \alpha_2 \wedge \alpha_4, \\ d\alpha_6 = \alpha_1 \wedge \alpha_4 + \alpha_2 \wedge \alpha_3; \end{cases}$$

$$n_{6,18} = \begin{cases} d\alpha_4 = \alpha_1 \wedge \alpha_2, \\ d\alpha_5 = \alpha_1 \wedge \alpha_3, \\ d\alpha_6 = \alpha_2 \wedge \alpha_4; \end{cases}$$

$$n_{6,19} = \begin{cases} d\alpha_4 = \alpha_1 \wedge \alpha_2, \\ d\alpha_5 = \alpha_1 \wedge \alpha_3, \\ d\alpha_6 = \alpha_1 \wedge \alpha_4 + \alpha_2 \wedge \alpha_3; \end{cases}$$

$$n_{6,20} = \begin{cases} d\alpha_4 = \alpha_1 \wedge \alpha_2, \\ d\alpha_5 = \alpha_1 \wedge \alpha_3, \\ d\alpha_6 = \alpha_1 \wedge \alpha_4; \end{cases}$$

$$n_{6,21} = \begin{cases} d\alpha_4 = \alpha_1 \wedge \alpha_2, \\ d\alpha_5 = \alpha_1 \wedge \alpha_3, \\ d\alpha_6 = \alpha_2 \wedge \alpha_3; \end{cases}$$

$$n_{6,23} = \begin{cases} d\alpha_5 = \alpha_1 \wedge \alpha_2 - \alpha_3 \wedge \alpha_4, \\ d\alpha_6 = \alpha_1 \wedge \alpha_3 + \alpha_2 \wedge \alpha_4; \end{cases}$$

$$n_{6,24} = \begin{cases} d\alpha_5 = \alpha_1 \wedge \alpha_2, \\ d\alpha_6 = \alpha_1 \wedge \alpha_3 + \alpha_2 \wedge \alpha_4; \end{cases}$$

$$n_{5,1} \oplus \mathbb{K} = \begin{cases} d\alpha_3 = \alpha_1 \wedge \alpha_2, \\ d\alpha_4 = \alpha_1 \wedge \alpha_3, \\ d\alpha_5 = \alpha_1 \wedge \alpha_4 + \alpha_2 \wedge \alpha_3; \end{cases}$$

$$n_{5,2} \oplus \mathbb{K} = \begin{cases} d\alpha_3 = \alpha_1 \wedge \alpha_2, \\ d\alpha_4 = \alpha_1 \wedge \alpha_3, \\ d\alpha_5 = \alpha_1 \wedge \alpha_4; \end{cases}$$

$$n_{5,4} \oplus \mathbb{K} = \begin{cases} d\alpha_4 = \alpha_1 \wedge \alpha_2, \\ d\alpha_5 = \alpha_2 \wedge \alpha_3 + \alpha_1 \wedge \alpha_4; \end{cases}$$

$$n_{5,5} \oplus \mathbb{K} = \begin{cases} d\alpha_4 = \alpha_1 \wedge \alpha_2, \\ d\alpha_5 = \alpha_1 \wedge \alpha_3; \end{cases}$$

$$n_{4,1} \oplus \mathbb{K}^2 = \begin{cases} d\alpha_3 = \alpha_1 \wedge \alpha_4, \\ d\alpha_2 = \alpha_1 \wedge \alpha_3; \end{cases}$$

$$n_3 \oplus n_3 = \begin{cases} d\alpha_3 = \alpha_1 \wedge \alpha_2, \\ d\alpha_6 = \alpha_4 \wedge \alpha_5; \end{cases}$$

$$n_3 \oplus \mathbb{K}^3 = \{ d\alpha_3 = \alpha_1 \wedge \alpha_2;$$

$$a_6 = \text{abelian Lie algebra}$$

The above notations correspond to classifications given in [33]. The trivial equations of structure ($d\alpha = 0$) have been omitted in the previous definitions.

6.6.3 Exact k-symplectic Lie algebras

Let $\mathcal{A}$ be a k-symplectic Lie algebra defined by

$$\left(\theta^1, \ldots, \theta^k; \mathcal{J} \right).$$

In addition we will assume that $\mathcal{J}$ is an ideal of $\mathcal{A}$ of codimension n and that the forms defining the k-symplectic structure are exact

$$\theta^1 = d\omega^1, \ldots, \theta^k = d\omega^k,$$

where $\omega^1, \ldots, \omega^k$ are independent linear forms on the Lie algebra $\mathcal{A}$ satisfying

$$\mathcal{A} = \mathcal{J} + (\ker \omega^1 \cap \cdots \cap \ker \omega^k).$$

These Lie algebras will be called *exact of type* $(k, n; \mathcal{J})$.

Lemma 22 *In the previous hypothesis and notations, there exists a basis* $(X_j)_{1\leq j\leq nk}$ *of* $\mathcal{J}$ *such that*

$$\omega^r(X_j) = \delta_j^r$$

for all r, j ($r = 1, \ldots, k$ *and* $j = 1, \ldots, nk$).

Proof. Let $e_1, \ldots, e_k$ be vectors belonging to $\mathcal{A}$ such that $\omega^r(e_j) = \delta_j^r$. The above hypothesis proves that for any i ($i = 1, \ldots, k$) the vectors e_i can be written in the form $e_i = X_i + Y_i$, where $X_i \in \mathcal{J}$ and $Y_i \in \ker\omega^1 \cap \cdots \cap \ker\omega^k$. For each i($i = 1, \ldots, k$) we have

$$\omega^r(X_i) = \omega^r(e_i) = \delta_i^r.$$

The vectors $X_1, \ldots, X_k$ are linearly independent. We show first that we can extend this system to a basis $(X_1, \ldots, X_k, X'_{k+1}, \ldots, X'_{nk})$ of $\mathcal{J}$ such that $\omega^r(X'_u)$ is not zero for $r = 1, \ldots, k$ and $u > k$.

Let $(X_1, \ldots, X_k, X''_{k+1}, \ldots, X''_{nk})$ be a basis $\mathcal{J}$ containing such a system. For all $r = 1, \ldots, k$ and $u = k + 1, \ldots, nk$ we take

$$\rho_u^r = \omega^r(X_u).$$

Let γ be a real number not belonging to the set

$$\{\rho_u^r \mid r = 1, \ldots, k, \ u = k + 1, \ldots, nk\} \cap \mathbb{K}^*.$$

We define

$$X'_u = X''_u - \gamma(X_1 + \ldots + X_k).$$

The vectors $X_1, \ldots, X_k, X'_{k+1}, \ldots, X'_{nk}$ form a basis $\mathcal{J}$ such that

$$\omega^r(X'_u) = \rho_u^r - \gamma \neq 0$$

for all $r = 1, \ldots, k$ and $u = k + 1, \ldots, nk$. For all $A = (a_i^j) \in \mathrm{GL}(n - 1, \mathbb{K})$ we take

$$X_{k+v} = -\sum_{p=1}^{k} \sum_{j=1}^{(n-1)k} a_v^j \omega^p(X'_{k+j}) X_p - \sum_{j=1}^{(n-1)k} a_v^j X'_{k+j}$$

for every $v = 1, \ldots, n - 1$. The vectors $(X_j)_{1\leq j\leq nk}$ form a basis of $\mathcal{J}$ such that $\omega^r(X_j) = \delta_j^r$ for all $r = 1, \ldots, k$ and $j = 1, \ldots, nk$. ∎

Proposition 6.9 *If $(\theta^1 = d\omega^1, \ldots, \theta^k = d\omega^k; \mathcal{J})$ is an exact k-symplectic structure on $\mathcal{A}$, then we can extend $(\omega^1, \ldots, \omega^k)$ to a basis*

$$(\omega^1, \ldots, \omega^k, \omega^{k+1}, \ldots, \omega^{n(k+1)})$$

of $\mathcal{A}^*$ such that the Lie algebra law of $\mathcal{A}$ is given by

$$d\omega^p = \sum_{a=nk+1}^{n(k+1)} \left(\sum_{s<a} C_{sa}^p \omega^s \right) \wedge \omega^a, \quad p = 1, \ldots, k,$$

$$d\omega^r = \sum_{nk+1 \leq u < v \leq n(k+1)} C_{uv}^r \omega^u \wedge \omega^v, \quad r = nk+1, \ldots, n(k+1),$$

and the matrix

$$\begin{pmatrix}
C^1_{1,nk+1} & C^1_{2,nk+1} & \cdots & C^1_{nk,nk+1} \\
\vdots & \vdots & & \vdots \\
C^r_{1,nk+1} & C^r_{2,nk+1} & \cdots & C^r_{nk,nk+1} \\
\vdots & \vdots & & \vdots \\
C^r_{1,n(k+1)} & C^r_{2,n(k+1)} & \cdots & C^r_{nk,n(k+1)} \\
\vdots & \vdots & & \vdots \\
C^k_{1,nk+1} & C^k_{2,nk+1} & \cdots & C^k_{nk,nk+1} \\
\vdots & \vdots & & \vdots \\
C^k_{1,n(k+1)} & C^k_{2,n(k+1)} & \cdots & C^k_{nk,n(k+1)}
\end{pmatrix}$$

is invertible.

Proof. Let $(X_j)_{1 \leq j \leq nk}$ be a basis of $\mathcal{J}$ such that

$$\omega^r(X_j) = \delta_j^r$$

for all r, j ($r = 1, \ldots, k$ and $j = 1, \ldots, nk$), let $(X_{nk+1}, \ldots X_{n(k+1)})$ be a basis of $\ker \omega^1 \cap \cdots \cap \ker \omega^k$, and let $(\omega'^1, \ldots, \omega'^k, \omega^{k+1}, \ldots, \omega^{n(k+1)})$ be the dual basis of $(X_1, \ldots, X_k, X_{k+1}, \ldots, X_{n(k+1)})$. The relationship

$$\omega^r(X_j) = \omega'^r(X_j) = \delta_j^r$$

for all $r, j = 1, \ldots, k$ proves that $\omega^p = \omega'^p$ for every $p = 1, \ldots, k$. But $\mathcal{J}$ is an ideal of $\mathcal{A}$, thus

$$[X_s, X_a] = - \sum_{r=1}^{n(k+1)} C_{sa}^r X_r \in \mathcal{J}$$

for all $s = 1, \ldots, nk$ and $a = 1, \ldots, n(k+1)$, consequently we have

$$C^r_{sa} = 0$$

for all $r = nk+1, \ldots, n(k+1)$, $s = 1, \ldots, nk$; this proves that

$$dw^r = \sum_{nk+1 \leq u < v \leq n(k+1)} C^r_{uv} w^u \wedge w^v,$$

where $r = nk+1, \ldots, n(k+1)$. For $r = 1, \ldots, k$, we write

$$dw^r = \sum_{nk+1 \leq s < j \leq nk} C^r_{sj} w^s \wedge w^j + \sum_{a=nk+1}^{n(k+1)} (\sum_{s<a} C^r_{sa} w^s) \wedge w^a.$$

The second condition of the definition of a k-symplectic structure implies that

$$C^r_{sj} = 0$$

for all $r = 1, \ldots, k$ and $s, j = 1, \ldots, nk$, thus

$$dw^r = \sum_{a=nk+1}^{n(k+1)} (\sum_{s<a} C^r_{sa} w^s) \wedge w^a.$$

The non-degeneracy of the k-symplectic system $\{\theta^1, \theta^2, \cdots, \theta^k\}$ proves that the matrix C is invertible. ■

The exact 1-codimensional k-symplectic Lie algebras (that is, $n = 1$) are completely determined by :

Corollary 6.3 *In the previous hypothesis and notations, if $n = 1$ the law of Lie algebras of $\mathcal{A}$ is given by:*

$$\begin{cases} dw^1 = (a^1_1 w^1 + \ldots + a^1_k w^k) \wedge w^{k+1}, \\ \qquad\qquad \vdots \\ dw^k = (a^k_1 w^1 + \ldots + a^k_k w^k) \wedge w^{k+1}, \\ dw^{k+1} = 0, \end{cases}$$

where $A = (a^u_v)_{1 \leq u,v \leq k}$ is invertible.

6.6.4 Lie algebras of the type $(2, 2; \mathcal{J})$

We consider here the case of 2-symplectic Lie algebras of the type $(2, 2; \mathcal{J})$. These Lie algebras are given by:

$$d\omega^p = \left(\sum_{s=1}^{4} C_{s5}^p \omega^s\right) \wedge \omega^5 + \left(\sum_{s=1}^{5} C_{s6}^p \omega^s\right) \wedge \omega^6, \quad p = 1, 2$$
$$d\omega^5 = u\omega^5 \wedge \omega^6,$$
$$d\omega^6 = v\omega^5 \wedge \omega^6,$$

where $u = C_{56}^5$ and $v = C_{56}^6$. The Jacobi equations $dd\omega^a = 0$ allow us to determine $d\omega^3$ and $d\omega^4$. In addition, we assume that we have

$$C_{15}^1 = C_{25}^2 = C_{36}^1 = C_{46}^2 = 1,$$

the other structural constants C_{sj}^1 and C_{sj}^2 are zero. The Lie algebra for which all the other undefined structural constants are zero is called the 2-symplectic Lie algebra model (all 2-symplectic algebra is a deformation of this model). From the relations $dd\omega^1 = dd\omega^2 = 0$ we have

$$\begin{cases} C_{35}^3 = 1 + v, \\ C_{15}^3 = u, \\ C_{12}^3 = C_{13}^3 = C_{14}^3 = C_{23}^3 = C_{24}^3 = C_{25}^3 = C_{34}^3 = C_{45}^3 = 0, \\ C_{45}^4 = 1 + v, \\ C_{25}^4 = u, \\ C_{12}^4 = C_{13}^4 = C_{14}^4 = C_{15}^4 = C_{23}^4 = C_{24}^4 = C_{34}^4 = C_{35}^4 = 0. \end{cases}$$

The relations $dd\omega^3 = dd\omega^4 = 0$ imply

$$2u + v(C_{36}^3 + u) = 0,$$
$$-u^2 - 2vC_{16}^3 + uC_{36}^3 = 0,$$
$$-2vC_{26}^3 + uC_{46}^3 = 0,$$
$$vC_{46}^3 = 0,$$
$$-2vC_{16}^4 + uC_{36}^4 = 0,$$
$$2u + uv + vC_{46}^4 = 0,$$
$$-u^2 - 2vC_{26}^4 + uC_{46}^4 = 0,$$
$$vC_{36}^4 = 0.$$

If $v = 0$, then we have $u = 0$; thus we obtain a family of such Lie algebras depending on 10 parameters for which structural equations are

$$
\begin{cases}
d\omega^1 = \omega^1 \wedge \omega^5 + \omega^3 \wedge \omega^6, \\
d\omega^2 = \omega^2 \wedge \omega^5 + \omega^4 \wedge \omega^6, \\
d\omega^3 = \omega^3 \wedge \omega^5 + C_{16}^3 \omega^1 \wedge \omega^6 + C_{26}^3 \omega^2 \wedge \omega^6 + C_{36}^3 \omega^3 \wedge \omega^6 + C_{46}^3 \omega^4 \wedge \omega^6 \\
\qquad + C_{56}^3 \omega^5 \wedge \omega^6, \\
d\omega^4 = C_{16}^4 \omega^1 \wedge \omega^6 + C_{26}^4 \omega^2 \wedge \omega^6 + C_{36}^4 \omega^3 \wedge \omega^6 + \omega^4 \wedge \omega^5 + C_{46}^4 \omega^4 \wedge \omega^6 \\
\qquad + C_{56}^4 \omega^5 \wedge \omega^6, \\
d\omega^5 = 0, \\
d\omega^6 = 0.
\end{cases}
$$

When $v \neq 0$, we obtain

$$
\begin{aligned}
C_{46}^3 = C_{36}^4 = C_{26}^3 = C_{16}^4 &= 0, \\
C_{26}^4 = C_{16}^3 &= -\frac{u^2(1+v)}{v^2}, \\
C_{36}^3 = C_{46}^4 &= -\frac{u(2+v)}{v}.
\end{aligned}
$$

The constants C_{56}^3 and C_{56}^4 being arbitrary; thus we obtain a family of such Lie algebras depending on 4 parameters forwhichstructuralequationsare

$$
\begin{cases}
d\omega^1 = \omega^1 \wedge \omega^5 + \omega^3 \wedge \omega^6, \\
d\omega^2 = \omega^2 \wedge \omega^5 + \omega^4 \wedge \omega^6, \\
d\omega^3 = u\omega^1 \wedge \omega^5 - \frac{u^2(1+v)}{v^2}\omega^1 \wedge \omega^6 + (1+v)\omega^3 \wedge \omega^5 - \frac{u(2+v)}{v}\omega^3 \wedge \omega^6. \\
\qquad + C_{56}^3 \omega^5 \wedge \omega^6, \\
d\omega^4 = u\omega^2 \wedge \omega^5 - \frac{u^2(1+v)}{v^2}\omega^2 \wedge \omega^6 + (1+v)\omega^4 \wedge \omega^5 - \frac{u(2+v)}{v}\omega^4 \wedge \omega^6 \\
\qquad + C_{56}^4 \omega^5 \wedge \omega^6, \\
d\omega^5 = u\omega^5 \wedge \omega^6, \\
d\omega^6 = v\omega^5 \wedge \omega^6.
\end{cases}
$$

Chapter 7

$k-$SYMPLECTIC AFFINE MANIFOLDS

7.1   Affine manifolds

Let M be a smooth manifold of dimension n. We say that M is an affine manifold if there is an atlas (U_i, φ_i) of M such that the changes of coordinates are restrictions of affine transformations of $\mathbb{R}^n$. An affine structure on M is equivalent to a given connection

$$\nabla : \Gamma(TM) \times \Gamma(TM) \longrightarrow \Gamma(TM)$$

such that both the curvature

$$k(X,Y) = \nabla_{[X,Y]} - (\nabla_X \nabla_Y - \nabla_Y \nabla_X)$$

and torsion

$$T(X,Y) = \nabla_X Y - \nabla_Y X - [X,Y]$$

vanish identically.

Example 23 *The circle S^1 is an affine manifold. It is, in fact, the unique sphere equipped with an affine structure.*

7.2   Affine Lie group

7.2.1   Definition

A Lie group G is called affine, if there exists a left-invariant affine structure on G, that is, for every $x \in G$ the mapping

$$L_x : G \to G$$

defined by

$$L_x(y) = xy$$

is affine. In other words, the group G is endowed with a left-invariant flat affine connection.

7.2.2   Left symmetric product

Let $\mathcal{G}$ be the Lie algebra of the affine Lie group G. By identifying the left-invariant vector fields on G with the elements of $\mathcal{G}$ we define a new product on $\mathcal{G}$ by taking

$$X \cdot Y = \nabla_X Y.$$

The property that ∇ is without curvature, implies that this product verifies

$$X \cdot (Y \cdot Z) - Y \cdot (X \cdot Z) = [X,Y] \cdot Z.$$

Such a product is termed *left-symmetric*.

Each symplectic Lie group is affine. In fact, the product in the Lie algebra $\mathcal{G}$ given by:

$$\theta(X \cdot Y, Z) = -\theta(Y, [X, Z])$$

is left symmetric associated with the bracket, where θ is a left-invariant symplectic form.

Proposition 7.1 *Every 1-symplectic Lie group is affine.*

In this chapter, we are going to show that every k−symplectic manifold is affine with respect to the differential structure defined by the leaves underlying the k−symplectic structure.

7.3 Characteristic foliations

7.3.1 Definition

Let M be an $n(k+1)$-dimensional smooth manifold. Let E be an integrable n-codimensional sub-bundle of TM. Let $\theta^1, \ldots, \theta^k$ be closed 2-forms on M. Let TM/E be the quotient bundle. Let ν be the canonical projection $TM \longrightarrow TM/E = \nu E$. Let $\nu^* E$ be the dual of νE. We denote by $\mathfrak{F}$ the foliation defined by the sub-bundle E. For each p ($p = 1, \ldots, k$) and $x \in M$ we denote by $C_x(\theta^p)$ the characteristic space of θ^p at x. All geometric objects on M are assumed to be of class C^∞.

Let $(\theta^1, \ldots, \theta^k; E)$ be a k-symplectic structure on M. Recall that about every point of M we can find an open neighborhood U and local coordinates $(x^{pi}, x^i)_{1 \leq p \leq k, 1 \leq i \leq n}$ on U, called adapted coordinates, such that

$$\theta^p_{|U} = \sum_{i=1}^{n} dx^{pi} \wedge dx^i, \qquad E_{|U} = \ker dx^1 \cap \ldots \cap \ker dx^n$$

for each p ($p = 1, \ldots, k$).

Equivalently, there exists an atlas $\mathfrak{A}$ of M, called the Darboux atlas, whose the changes of coordinates belong to the pseudogroup of local diffeomorphisms of $\mathbb{R}^{n(k+1)}$ leaving the canonical k-symplectic structure invariant.

It follows from the previous canonical Darboux expressions that the differential forms θ^p are of class $2n$; this implies that the distribution

$$x \longmapsto C_x(\theta^p)$$

defines an integrable sub-bundle of TM.

For each p ($p = 1, \ldots, k$) we take

$$E^p = \bigcap_{q \neq p} C(\theta^q).$$

In terms of the adapted coordinates $(x^{pi}, x^i)_{1 \leq p \leq k, 1 \leq i \leq n}$, we have :

1. νE is spanned by the derivatives $\partial/\partial x^1, \ldots, \partial/\partial x^n$;

2. $\nu_* E$ is spanned by the differential forms $dx^1, \ldots, dx^n$;

3. E^p is spanned by the derivatives $\partial/\partial x^{p1}, \ldots, \partial/\partial x^{pn}$.

Proposition 7.2 *We have:*

1. *For each p $(p = 1, \ldots, k)$ the sub-bundle E^p is integrable;*

2. *$E = E^1 \oplus \ldots \oplus E^k$ (direct sum) and $[E^p, E^q] \subset \cap_{r \neq p,q} C(\theta^r)$*

3. *For each p $(p = 1, \ldots, k)$ the mapping $X \longmapsto i(X)\theta^p$ defines an isomorphism i_p of vector bundles over M of E^p onto $\nu^* E$.*

Proof. Only the second part of the second assertion necessitates a comment. Let $X \in E^p$ and $Y \in E^q$. We have

$$
\begin{aligned}
i([X, Y])\theta^r &= L_X i(Y)\theta^r - i(Y)L_X\theta^r \\
&= -i(Y)L_X\theta^r \\
&= -i(Y)(i(X)d\theta^r + d(i(X)\theta^r)) \\
&= 0
\end{aligned}
$$

if $r \neq p$ and q. ∎

Definition 7.1 *The foliations $\mathfrak{F}^p$ of M defined by the sub-bundles E^p are called the characteristic foliations of the k-symplectic structure.*

In terms of the adapted coordinates $(x^{pi}, x^i)_{1 \leq p \leq k, 1 \leq i \leq n}$ the mappings i_p express the duality

$$
\frac{\partial}{\partial x^{ps}} \longmapsto dx^s
$$

between the geometry along the leaves of $\mathfrak{F}^p$ and the transverse geometry.

Proposition 7.3 *For every basic function f on M, we can associate k vector fields $X_f^1, \ldots, X_f^k$ satisfying*

1. *X_f^p is tangent to leaves of characteristic foliations $\mathfrak{F}^p$ for each p,*

2. *$\left[X_f^p, X_f^q\right] = 0$.*

Proof. By the duality $i_p : X \longmapsto i(X)\theta^p$ from E^p onto the transverse space $\nu^* E$ we can, for each smooth basic function f, associate k vector fields $X_f^1, \ldots, X_f^k$ such that

$$
i(X_f^p)\theta^p = -df
$$

for each p. With respect to the local coordinate system $(x^{pi}, x^i)_{1 \leq p \leq k, 1 \leq i \leq n}$ the vector field X_f^p can be written

$$X_f^p = -\sum_{i=1}^{n} \frac{\partial f}{\partial x^i} \frac{\partial}{\partial x^{pi}}.$$

It is clear that we have $X_f^p \in E^p$, and the first point is proved.

Let $p, q = 1, \ldots, k$. For every $r = 1, \ldots, k$, we have

$$\begin{aligned} L_{X_f^p} i(X_f^q)\theta^r &= \delta^{rq} \left(L_{X_f^p} df \right) \\ &= d \left(X_f^p(f) \right) \\ &= 0 \end{aligned}$$

and

$$\begin{aligned} i\left(\left[X_f^p, X_f^q\right]\right)\theta^r &= L_{X_f^p} i(X_f^q)\theta^r - i(X_f^q) L_{X_f^p}\theta^r \\ &= -i(X_f^q) L_{X_f^p}\theta^r \\ &= -i(X_f^q)\left(-i(X_f^p)d\theta^r + d\left(i(X_f^p)\theta^r\right)\right) \\ &= 0. \end{aligned}$$

Consequently we have

$$\left[X_f^p, X_f^q\right] \in \bigcap_{r=1}^{k} C\left(\theta^r\right) = \{0\}.$$

This completes the proof of the proposition. ∎

7.3.2 Affine structures on the leaves $\mathfrak{F}^p$

Let $\mathfrak{A}$ be the Darboux atlas given by the k-symplectic structure $(\theta^1, \ldots, \theta^k; E)$. The Darboux expressions imply the following local transition formulas of the canonical coordinates

$$\bar{x}^{pi} = \sum_{j=1}^{n} \frac{\partial x^j}{\partial \bar{x}^i} x^{pj} + \varphi^{pi}(x^1, \ldots, x^n), \qquad \bar{x}^i = \bar{x}^i(x^1, \ldots, x^n),$$

where φ^{pi} are some well defined basic functions. Indeed, these expressions are affine with respect to x^{pi}. Then we have

Proposition 7.4 *Each leaf of $\mathfrak{F}^p$ (resp. $\mathfrak{F}$) is locally affine.*

The affine structure of the leaves under the discussion has as local parallel vector fields those corresponding to the E-projectable 1-forms under the isomorphism i_p.

7.3.3   Associated connection

We now give a flat affine connection along the leaves. Let $\overline{\nabla}$ be a basic connection on νE, that is, an $\mathbb{R}$-bilinear mapping

$$\overline{\nabla} : \Gamma(E) \times \Gamma(\nu E) \longrightarrow \Gamma(\nu E)$$

such that

$$\overline{\nabla}_X Z = \nu[X, \overline{Z}] \, ,$$

for each $X \in \Gamma(E)$ and $Z \in \Gamma(\nu E)$, where $\overline{Z} \in \mathfrak{X}(M)$ is such that $\nu(\overline{Z}) = Z$.

Recall that the curvature $\bar{k}$ of $\overline{\nabla}$ is a correspondence which associates with every pair $(X, Y) \in \Gamma(E) \times \Gamma(E)$ a mapping

$$\bar{k}(X, Y) : \Gamma(\nu E) \longrightarrow \Gamma(\nu E)$$

given by

$$\bar{k}(X, Y)Z = \overline{\nabla}_X \overline{\nabla}_Y Z - \overline{\nabla}_Y \overline{\nabla}_X Z - \overline{\nabla}_{[X,Y]} Z, \quad Z \in \Gamma(\nu E).$$

We have $\bar{k}(X, Y) = 0$ for each $X, Y \in \Gamma(E)$.

Let $p = 1, \ldots, k$ and let $\mathfrak{L}$ be a leaf of $\mathfrak{F}^p$. For each $X, Y \in \mathfrak{X}(\mathfrak{L})$ we take

$$\nabla_X Y = i_p^{-1}(\overline{\nabla}^*_X(i_p(Y))),$$

where $\overline{\nabla}^*$ is the map of

$$\Gamma(E^p) \times \Gamma(\nu^* E) \longrightarrow \Gamma(\nu^* E)$$

defined by

$$\left(\overline{\nabla}^*_X \sigma\right) Z = L_X(\sigma(Z)) - \sigma(\overline{\nabla}_X Z)$$

for each $X \in \Gamma(E^p)$, $\sigma \in \Gamma(\nu^* E)$ and $Z \in \Gamma(\nu E)$ and L_X denote the Lie derivative with respect to X. ∇ defines an affine connection along the leaf $\mathfrak{L}$. Let T be the torsion tensor of ∇. Let Z be a local cross-section of the bundle $\nu E \longrightarrow M$ and let $\overline{Z}$ be a local cross-section of the bundle $TM \longrightarrow M$ such that $\nu(\overline{Z}) = Z$. We have

$$\theta^p(T(X, Y), \overline{Z}) = \theta^p(\nabla_X Y - \nabla_Y X, \overline{Z}) - \theta^p([X, Y], \overline{Z})$$

and

$$\begin{aligned}
\theta^p(\nabla_X Y - \nabla_Y X, \overline{Z}) &= \theta^p(i_p^{-1}(\overline{\nabla}^*_X(i_p(Y)) - \overline{\nabla}^*_Y(i_p(X))), \overline{Z}) \\
&= (\overline{\nabla}^*_X(i_p(Y)))(Z) - (\overline{\nabla}^*_Y(i_p(X)))(Z),
\end{aligned}$$

then

$$
\begin{aligned}
\theta^p(\nabla_X Y - \nabla_Y X, \overline{Z}) &= L_X i_p(Y)(Z) - i_p(Y)\overline{\nabla}_X Z - L_Y i_p(X)(Z) \\
&\quad + i_p(X)\overline{\nabla}_Y Z, \\
&= L_X(\theta^p(Y,\overline{Z})) - \theta^p(Y,[X,\overline{Z}]) \\
&\quad + L_Y(\theta^p(\overline{Z},X)) + \theta^p(X,[Y,\overline{Z}]).
\end{aligned}
$$

Consequently, we have

$$
\begin{aligned}
\theta^p(T(X,Y),\overline{Z}) &= L_X(\theta^p(Y,\overline{Z})) + L_Y(\theta^p(\overline{Z},X)) \\
&\quad + \theta^p(X,[Y,\overline{Z}]) - \theta^p(Y,[X,\overline{Z}]) - \theta^p([X,Y],\overline{Z}), \\
&= d\theta^p(X,Y,\overline{Z}).
\end{aligned}
$$

As θ^p is closed, Z arbitrary and i_p is an isomorphism, then

$$
T(X,Y) = 0.
$$

The curvatures k, $\bar{k}$ and $\bar{k}^*$ of ∇, $\overline{\nabla}$ and $\overline{\nabla}^*$ respectively are related by

$$
i_p(k(X_1,X_2)Y) = \bar{k}^*(X_1,X_2)(i_p(Y))
$$

where X_1, X_2, $Y \in \mathfrak{X}(\mathcal{L})$. We have $\bar{k} \equiv 0$, thus $\bar{k}^* \equiv 0$ and $k \equiv 0$ and we have proved the following proposition:

Proposition 7.5 ∇ *is a flat connection; it defines an affine structure along the leaf* $\mathcal{L}$.

We are now going to define a flat connection on the leaves of the foliation $\mathfrak{F}$ inducing other flat affine connections. Let $\mathfrak{M}$ be a leaf of $\mathfrak{F}$. For every $X, Y \in \mathfrak{X}(\mathfrak{M})$ we take

$$
D_X Y = \sum_{p=1}^{k} i_p^{-1}(\overline{\nabla}_X^*(i_p(Y^p))),
$$

where $Y^p \in \Gamma(E^p)$ such that $Y = Y^1 + \cdots + Y^k$. It is clear that D defines a connection on $\mathfrak{M}$ whose restriction to the leaf $\mathcal{L}$ coincides with ∇. This connection is without torsion. In fact, if X and Y belong to $\Gamma(E)$, then

$$
\theta^p(T_1(X,Y),\overline{Z}) = \theta^p(D_X Y - D_Y X, \overline{Z}) - \theta^p([X,Y],\overline{Z})
$$

and

$$\theta^p(D_X Y - D_Y X, \overline{Z}) = \sum_{q=1}^{k} \theta^p(i_q^{-1}(\overrightarrow{\nabla}_X^*(i_q(Y^q)) - \overrightarrow{\nabla}_Y^*(i_q(X^q))), \overline{Z})$$
$$= \theta^p(i_p^{-1}(\overrightarrow{\nabla}_X^*(i_p(Y^p)) - \overrightarrow{\nabla}_Y^*(i_p(X^p))), \overline{Z}).$$

Thus

$$\theta^p(D_X Y - D_Y X, \overline{Z}) = L_X i_p(Y^p)(Z) - i_p(Y^p)\overline{\nabla}_X Z - L_Y i_p(X^p)(Z)$$
$$+ i_p(X^p)\overline{\nabla}_Y Z,$$
$$= L_X(\theta^p(Y^p, \overline{Z})) - \theta^p(Y, [X, \overline{Z}])$$
$$+ L_Y(\theta^p(\overline{Z}, X^p)) + \theta^p(X, [Y, \overline{Z}]),$$

and

$$\theta^p(T_1(X, Y), \overline{Z}) = L_X(\theta^p(Y, \overline{Z})) + L_Y(\theta^p(\overline{Z}, X))$$
$$+ \theta^p(X, [Y, \overline{Z}]) - \theta^p(Y, [X, \overline{Z}]) - \theta^p([X, Y], \overline{Z})$$
$$= d\theta^p(X, Y, \overline{Z})$$
$$= 0.$$

Therefore the torsion T_1 of D is zero. This proves, in particular, that if $X \in \Gamma(E^p)$ and $Y \in \Gamma(E^q)$ then

$$[X, Y] = D_X Y - D_Y X.$$

As for the vanishing of the curvature, it is proved analogously.

Theorem 7.1 *Let M be a k-symplectic manifold. Then the leaves of $\mathfrak{F}$ and $\mathfrak{F}^p$ are affine.*

7.3.4   Consequences

Recall the Blumenthal theorem [11]

Proposition 7.6 *Let M be a smooth compact manifold with finite fundamental group. Then M does not support a foliation with a flat basic connection.*

Corollary 7.1 *Let $(M, \mathcal{F})$ be a foliatedd manifold with a flat basic connection. If $H_1(M, \mathbb{Z}) = 0$ then the foliation $\mathfrak{F}$ admits a transverse volume element; that is, $\mathfrak{F}$ is defined by a nowhere zero closed q-form on M where q is the codimension of $\mathfrak{F}$.*

As consequence we have:

Proposition 7.7 *Let M be an $n(k+1)$-dimensional manifold equipped with a k-symplectic structure. Then:*

1. *If M is compact then the fundamental group $\pi_1(M)$ of M is infinite;*

2. *If $H_1(M,\mathbb{Z}) = 0$ then the foliation $\mathfrak{F}$ is defined by a nowhere zero closed n-form on M.*

Recall that a foliated vector field X of the sub-bundle E is a Hamiltonian system if the differential forms $i(X)\theta^1,\ldots,i(X)\theta^k$ are closed.

Let $H = (H^p)_{1\leq p\leq k}$ be a Hamiltonian mapping and let X be the associated Hamiltonian system. With respect to the adapted coordinate system $(x^{pi},x^i)_{1\leq p\leq k,1\leq i\leq n}$, the components H^p of H and X_H can be written:

$$H^p = \sum_{j=1}^{n} f_j(x^1,\ldots,x^n)x^{pj} + g^p(x^1,\ldots,x^n),$$

and the vector field X takes the form

$$
X = \sum_{i=1}^{n}\left(-\sum_{p=1}^{k}\left(\sum_{j=1}^{n}\frac{\partial f_j}{\partial x^i}(x^1,\ldots,x^n)x^{pi} + \frac{\partial g^p}{\partial x^i}(x^1,\ldots,x^n)\right)\frac{\partial}{\partial x^{pi}}\right.
$$
$$
\left. + f_i(x^1,\ldots,x^n)\frac{\partial}{\partial x^i}\right)
$$

where f_j and g^p are smooth basic functions on U. Consequently we have:

Proposition 7.8 *The flow generated by X respects the affine structure of the leaves of $\mathfrak{F}$, that is, X is an infinitesimal affine transformation of the space M with its differential structure of dimension nk defined by the leaves of $\mathfrak{F}$ and the restriction of H to each leaf of $\mathfrak{F}$ is an affine transformation.*

7.4   Flag connection on a pair of sub-bundles

Let M be a smooth connected manifold and E be a q-dimensional sub-bundle of TM. Let $\Lambda^p(M)$ be the space of all p-forms on M and $\Gamma(E)$ (resp. $\mathfrak{X}(M)$) be the $\Lambda^0(M)$-module of all cross-sections of E (resp. TM). Let TM/E be the quotient bundle. Let ν be the canonical projection

$$TM \longrightarrow TM/E = \nu E.$$

Let ν^*E be the dual of νE. We denote by $\mathfrak{F}$ the foliation defined by E when this sub-bundle is integrable.

All geometric objects on M are assumed to be of class C^∞.

A transverse frame at a point x of M is a linear isomorphism of $\mathbb{R}^q$ onto the transverse space $(\nu E)_x$, and the set of these transverse frames is a fibre bundle $E_T(M, \nu_T, GL(q, \mathbb{R}))$ over M. Recall that a connection ω on E_T is said to be basic if the parallel translation which it defines along paths lying in a leaf of $\mathfrak{F}$ agrees with the 'natural parallelism along the leaves'. Equivalently, the associated Koszul operator

$$\nabla : \mathfrak{X}(M) \times \Gamma(\nu E) \longrightarrow \Gamma(\nu E)$$

satisfies the condition that

$$\nabla_X Z = \nu[X, \overline{Z}]$$

for all $X \in \Gamma(E)$ and all $Z \in \Gamma(\nu E)$, where $\overline{Z}$ is any vector field on M, such that $\nu(\overline{Z}) = Z$. Choosing of Riemannian metric on M, we may regard νE as a sub-bundle of TM complementary to E $(TM = E \oplus \nu E)$. If we denote by ∇ the covariant differentiation operator corresponding to the basic connection, then we have

$$\nabla_X Y = \nu[X, Y]$$

for all $X \in \Gamma(E)$ and all $Y \in \Gamma(\nu E)$.

Definition 7.2 *A basic connection on the sub-bundle E is an $\mathbb{R}$-bilinear map*

$$\overline{\nabla} : \Gamma(E) \times \Gamma(\nu E) \longrightarrow \Gamma(\nu E)$$

satisfying

$$\overline{\nabla}_X Z = \nu[X, \overline{Z}]$$

for each $X \in \Gamma(E)$ and $Z \in \Gamma(\nu E)$ where $\overline{Z} \in \mathfrak{X}(M)$ such that $\nu(\overline{Z}) = Z$.

For every $X \in \Gamma(E)$, $Y \in \Gamma(\nu E)$ and $f \in \Lambda^0(M)$ we have:

$$\overline{\nabla}_{fX} Y = f\overline{\nabla}_X Y, \quad \overline{\nabla}_X fY = X(f)Y + f\overline{\nabla}_X Y.$$

A linear connection π on $E_T(M, \nu_T, GL(q, \mathbb{R}))$ is basic if and only if the covariant differentiation operator corresponding to π gives a basic connection on the sub-bundle E. Using the basic connection on an integrable Lagrangian sub-bundle, we obtain:

Proposition 7.9 *Let (M,Ω) be a symplectic manifold equipped with a Lagrangian foliation $\mathfrak{F}$. Then each leaf of $\mathfrak{F}$ has a canonical flat affine connection.*

Let M be a smooth connected manifold, let E be a q-codimensional sub-bundle of TM and let F be a sub-bundle of TM contained in E ($F \subseteq E$). If E is integrable then

$$[X,Y] \in \Gamma(E)$$

for each X, $Y \in \Gamma(F)$. For $Z \in \Gamma(\nu E)$ these holds $Z = \nu(\overline{Z})$ for some $\overline{Z} \in \mathfrak{X}(M)$; $\overline{Z}$ is well defined modulo $\Gamma(E)$. Thus for $X \in \Gamma(F)$ and $Z \in \Gamma(\nu E)$

$$\overline{\nabla}_X Z = \nu[X,\overline{Z}]$$

is well defined. This is clearly $\mathbb{R}$-bilinear as a map

$$\overline{\nabla} : \Gamma(F) \times \Gamma(\nu E) \longrightarrow \Gamma(\nu E)$$

as can be verified

$$\overline{\nabla}_{fX} Y = f\overline{\nabla}_X Y, \quad \overline{\nabla}_X fY = X(f)Y + f\overline{\nabla}_X Y,$$

for all $X \in \Gamma(F)$, $Y \in \Gamma(\nu E)$, and for all $f \in \Lambda^0(M)$. This satisfies the definition of a connection on the sub-bundle E, except that the variable X is restricted to ranging over $\Gamma(F)$.

Definition 7.3 *A flag connection on the pair (E,F) is an $\mathbb{R}$-bilinear map*

$$\overline{\nabla} : \Gamma(F) \times \Gamma(\nu E) \longrightarrow \Gamma(\nu E)$$

such that

$$\overline{\nabla}_X Z = \nu[X,\overline{Z}],$$

for each $X \in \Gamma(F)$ and $Z \in \Gamma(\nu E)$, where $\overline{Z} \in \mathfrak{X}(M)$ satisfies $\nu(\overline{Z}) = Z$.

Such a connection will be called an F-connection on E. Notice that we have :

1. Under the assumption that the sub-bundle E admits a basic connection (in particular if E is integrable), there exists an F-connection on E, for each sub-bundle $F \subseteq E$.

2. A flag connection on the pair (E,E) is a basic connection on E.

Proposition 7.10 *Let $E = F^1 \oplus \ldots \oplus F^q$ be a direct sum of sub-bundles $F^1, \ldots, F^q$ of TM such that for each p we have an F^p-connection $\overline{\nabla}^p$ on E. Then the sub-bundle E admits a basic connection..*

Proof. Let $X = \sum_{p=1}^q X^p \in \Gamma(E)$, $X^p \in \Gamma(F^p)$ and let $Z \in \Gamma(\nu E)$. If we take

$$\overline{\nabla}_X Z = \sum_{p=1}^q \overline{\nabla}^p_{X^p} Z$$

we will obtain

$$\overline{\nabla}_X Z = \sum_{p=1}^q \nu[X^p, \overline{Z}] = \nu[X, \overline{Z}],$$

where $\overline{Z} \in \mathfrak{X}M)$ satisfies $\nu(\overline{Z}) = Z$; this proves that $\overline{\nabla}$ is a basic connection on E. ■

Proposition 7.11 *Let F be an integrable sub-bundle of TM, such that $F \subseteq E$, let $\overline{\nabla}$ be an F-connection on E and let $\bar{k}$ be its curvature. Then $\bar{k}(X, Y) = 0$ for each $X, Y \in \Gamma(F)$.*

Proof. The curvature $\bar{k}$ of the F-connection $\overline{\nabla}$ is a correspondence which associates with every pair $(X, Y) \in \Gamma(F) \times \Gamma(F)$ a mapping

$$\bar{k}(X, Y) : \Gamma(\nu E) \longrightarrow \Gamma(\nu E)$$

given by

$$\bar{k}(X, Y)Z = \overline{\nabla}_X \overline{\nabla}_Y Z - \overline{\nabla}_Y \overline{\nabla}_X Z - \overline{\nabla}_{[X,Y]} Z, \quad Z \in \Gamma(\nu E).$$

Let $X, Y \in \Gamma(F)$, $Z \in \Gamma(\nu E)$ and $\overline{Z} \in \mathfrak{X}(M)$ be such that $\nu(\overline{Z}) = Z$. If we take

$$\overline{\nu[X, \overline{Z}]} = [X, \overline{Z}], \quad \overline{\nu[Y, \overline{Z}]} = [Y, \overline{Z}],$$

we will obtain

$$\begin{aligned}
\bar{k}(X, Y)Z &= \overline{\nabla}_X(\nu[Y, \overline{Z}]) - \overline{\nabla}_Y(\nu[X, \overline{Z}]) - \nu[[X, Y], \overline{Z}] \\
&= \nu[X, [Y, \overline{Z}]] + \nu[Y, [\overline{Z}, X]] + \nu[\overline{Z}, [X, Y]] \\
&= 0. \blacksquare
\end{aligned}$$

The notion of flag connection allows us to find the existence of an affine structure on the leaves of the characteristic foliations.

7.5 k-symplectic affine manifolds

7.5.1 k-symplectic affine structure

Let M be an $n(k+1)$-dimensional manifold with a k-symplectic structure

$$(\theta^1, \ldots, \theta^k; E).$$

Definition 7.4 *We say that M is an affine k-symplectic manifold if the Darboux atlas $\mathfrak{A}$ confers upon M a structure of a locally affine manifold.*

Let $Gp(k, n; \mathbb{R})$ be the group of all affine transformations of $\mathbb{R}^{n(k+1)}$, leaving the canonical k-symplectic structure of $\mathbb{R}^{n(k+1)}$ invariant. The group $Gp(k, n; \mathbb{R})$ is the set of all affine transformations

$$X \longmapsto AX + B$$

of $\mathbb{R}^{n(k+1)}$ such that $A \in Sp(k, n; \mathbb{R})$.

Proposition 7.12 *Let M be a complete connected affine k-symplectic manifold of dimension $n(k+1)$. Then M is just a quotient $\mathbb{R}^{n(k+1)}/\Gamma$ and with a fundamental group Γ, where Γ is a subgroup of $A(n(k+1))$ acting freely and properly discontinuously on $\mathbb{R}^{n(k+1)}$:*

$$M = \mathbb{R}^{n(k+1)}/\Gamma, \quad \pi_1(M) = \Gamma.$$

Recall that a manifold M equipped with a connection ∇ is said to be complete if every geodesic of ∇ can be extended to a geodesic defined on $\mathbb{R}$.
Proof. M is a complete connected affine manifold, that is, a quotient $\mathbb{R}^{n(k+1)}/\Gamma$ whose fundamental group $\pi_1(M)$ is isomorphic to Γ where Γ is a subgroup of $A(n(k+1))$ acting freely and properly discontinuously on $\mathbb{R}^{n(k+1)}$

$$M = \mathbb{R}^{n(k+1)}/\Gamma, \quad \pi_1(M) = \Gamma.$$

Let $q : \mathbb{R}^{n(k+1)} \longrightarrow M = \mathbb{R}^{n(k+1)}/\Gamma$ be the canonical projection (q is the covering map), and let P be the sub-bundle of $T\mathbb{R}^{n(k+1)}$ defined by

$$q_* P_x = E_{q(x)}$$

for all $x \in \mathbb{R}^{n(k+1)}$. The $(k+1)$-tuple $(q^*\theta^1, \ldots, q^*\theta^1; P)$ defines a Γ-invariant k-symplectic structure on $\mathbb{R}^{n(k+1)}$.

The changes of coordinates of M are affine transformations of $\mathbb{R}^{n(k+1)}$ leaving the canonical k-symplectic structure invariant. By the definition of the quotient $M = \mathbb{R}^{n(k+1)}/\Gamma$ we see that Γ is contained in $Gp(k, n; \mathbb{R})$. ∎

7.5.2   Case where $\mathfrak{F}$ is 1-codimensional

Let $Hp(k,n;\mathbb{R})$ be the group of all matrices

$$\begin{pmatrix} I_n & & 0 & S_1 & T_1 \\ & \ddots & & \vdots & \vdots \\ 0 & & I_n & S_k & T_k \\ 0 & \cdots & 0 & I_n & Q \\ 0 & \cdots & 0 & 0 & 1 \end{pmatrix} \tag{7.1}$$

where I_n is the unit matrix of rank n, $S_1,\cdots,S_k$ are $n\times n$ real symmetric matrices and $T_1,\ldots,T_k$, Q are column vectors of length n. We denote by (S,Q,T) the matrices of the form (7.1), where $S=(S_1,\ldots,S_k)$, $T=(T_1,\ldots,T_k)$.

Proposition 7.13 *Let M be a complete connected affine k-symplectic manifold of dimension $k+1$. Then M is a quotient $\mathbb{R}^{k+1}/\Gamma$ and with a fundamental group Γ where Γ is a subgroup of $Hp(k,1;\mathbb{R})$ acting freely and properly discontinuously on $\mathbb{R}^{k+1}$*

$$M = \mathbb{R}^{k+1}/\Gamma, \qquad \pi_1(M) = \Gamma.$$

Proof.   It results from the previous proposition that M is a quotient $\mathbb{R}^{k+1}/\Gamma$ and has a fundamental group Γ where Γ is a subgroup of $A(k+1)$ acting freely and properly discontinuously on $\mathbb{R}^{k+1}$. An affine transformation $g \in Gp(k,1;\mathbb{R})$ has no a fixed point, if and only if the following conditions are satisfied:

1. $g = (S,q,T) \in Hp(k,1;\mathbb{R})$;

2. $T \notin \mathbb{R}.S$.

In particular we have $\Gamma \subset Hp(k,1;\mathbb{R})$. ∎

We assume now that $n=1$ and $k=2$. Let Γ_2^{02012} be the subgroup of $Hp(2,1;\mathbb{R})$ generated by

$$(S^0,0,T^0), \quad (mS^0,0,T^1), \quad (S^2,q_2,T^2),$$

where $S^0, S^2, T^0, T^1, T^2 \in \mathbb{Z}^2$, $q_2 \in \mathbb{N}^*$, $m \in \mathbb{Q}$ with

$$q_2 \neq 0, \quad \det(T^0,T^1) \neq 0 \quad \text{and} \quad \delta = mD_2$$

where δ and D_2 denote the determinants $\det(T^1,S^0)$ and $\det(T^0,S^0)$, respectively.

Proposition 7.14 *We have:*

1. *The subgroup* Γ_2^{02012} *acts freely and properly discontinuously on* $\mathbb{R}^3$.

2. *The quotient* $M_2^{02012} = \mathbb{R}^3/\Gamma_2^{02012}$ *is an orientable complete compact 2-symplectic connected locally affine manifold.*

3. *The manifold* M_2^{02012} *is a homeomorphic to the torus* T^3 *if and only if* $S^0 = 0$.

Proof.

1. Every element g of Γ_2^{02012} can be written $g = (S, q, T)$ with

(a') $\quad S = \left(\left(\sum_{i=1}^n \alpha_i\right) + m\left(\sum_{i=1}^n \beta_i\right)\right) S^0 + \left(\sum_{i=1}^n \gamma_i\right) S^2$,

(b') $\quad q = \left(\sum_{i=1}^n \gamma_i\right) q_2$,

(c') $\quad T = \left(\sum_{p=1}^n \left(\sum_{j=1}^p (\alpha_j + m\beta_j)\right) \gamma_p\right) q_2 S^0 + q_2 P_{\left(\sum_{i=1}^n \gamma_i\right)} S^2$
$\qquad\qquad + \left(\sum_{i=1}^n \alpha_i\right) T^0 + \left(\sum_{i=1}^n \beta_i\right) T^1 + \left(\sum_{i=1}^n \gamma_i\right) T^2$,

where $n \in \mathbb{N}$, $\alpha_i, \beta_i, \gamma_i \in \mathbb{Z}$ for all $i, j = 1, \ldots, n$ and $P_\gamma = \gamma(\gamma - 1)/2$.

If g admits a fixed point then $\sum_{i=1}^n \gamma_i = 0$ and there exists $\lambda \in \mathbb{Q}$ such that $T = \lambda S$; this proves that

$$\left(\sum_{j=1}^p (\alpha_j + m\beta_j)\right) = 0,$$

and also $g = (0, 0, 0)$. If $D_2 \neq 0$ then

$$\left(\sum_{j=1}^p (\alpha_j + m\beta_j)\right) D_2 = 0$$

and $g = (0, 0, 0)$. If $D_2 = 0$ then $\delta = 0$ and we have two possibilities

(a) if $S^0 = 0$ then $g = (0, 0, 0)$;

(b) if $S^0 \neq 0$ then $T^0, T^1 \in \mathbb{R}.S^0$ and $\det(T^0, T^1) = 0$; this contradicts the property $\det(T^0, T^1) \neq 0$. Thus $D_2 \neq 0$ and $g = (0, 0, 0)$.

Then the subgroup $G = \Gamma_2^{02012}$ acts freely on $\mathbb{R}^3$. We now show that G acts properly discontinuously on $\mathbb{R}^3$; that is, the following conditions are satisfied:

- Every point $Q_0 \in \mathbb{R}^3$ admits an open neighborhood U such that the set

$$\{g \in G \mid gU \cap U \neq \emptyset\}$$

is finite,

- If two points Q_1 and Q_2 of $\mathbb{R}^3$ are not congruent modulo G then Q_1 and Q_2 admit open neighborhoods U_1 and U_2 such that $gU_1 \cap U_2 = \emptyset$ for every $g \in G$.

Two cases are possible.
Case 1: $S^0 \neq 0$.

Let $p_0 \in \mathbb{Z}$ and $q_0 \in \mathbb{N}^*$ such that $m = p_0/q_0$. Let $Q_0 = (x_0, y_0, z_0) \in \mathbb{R}^3$, and

$$\varepsilon = \min(\frac{1}{2q_0}, \frac{|D_2|}{2q_0(|s_1^0| + |s_2^0|)}),$$

where $S^0 = (s_1^0, s_2^0)$, and let U_0 be the open ball $B(Q_0, \varepsilon)$ with center at the point Q_0 and with radius ε (for the norm $\|(u, v, w)\| = \max(|u|, |v|, |w|)$). Let $g \in G$ be such that $gU \cap U \neq \emptyset$ and let $Q = (x, y, z) \in U$ such that $g(Q) = (x', y', z') \in U$. Then we have

$$|x - x'| < 2\varepsilon,\, |y - y'| < 2\varepsilon \quad \text{and} \quad |z - z'| < 2\varepsilon.$$

By (a'), (b'), (c') we have

$$|z - z'| = |(\sum_{i=1}^{n} \gamma_i)q_2| < 2\varepsilon < 1,$$

then $|\sum_{i=1}^{n} \gamma_i|$ is an integer less than 1, $\sum_{i=1}^{n} \gamma_i = 0$ and $z = z'$. By the inequalities $|x - x'| < 2\varepsilon$ and $|y - y'| < 2\varepsilon$ we deduce that

$$|s_2^0(x' - x) - s_1^0(y' - y)| \leq 2\varepsilon(|s_1^0| + |s_2^0|).$$

And also by (a'), (b'), (c') we have

$$\left| \left((\sum_{i=1}^{n} \alpha_i) + m(\sum_{i=1}^{n} \beta_i) \right) q_0 \right| < 1,$$

then

$$\left(\sum_{i=1}^{n} \alpha_i\right) + m\left(\sum_{i=1}^{n} \beta_i\right) = 0.$$

It follows that $q_0|x - x'|$ and $q_0|y - y'|$ are two integers less than 1, then $x = x'$, $y = y'$ and also $g = (0,0,0)$; in other words we have

$$\{g \in G \mid gU \cap U \neq \emptyset\} = \{id_{\mathbb{R}^3}\}.$$

Consider now two points Q_1 and Q_2 of $\mathbb{R}^3$ not congruent modulo G; let K be a compact set contained in $\mathbb{R}^3$, and let $r > 0$ such that $K \subset B(Q_1, r)$, where $B(Q_1, r)$ is the open ball with center at the point Q_1 and with radius r.

The following sets

$$E_1 = \{\gamma \in \mathbb{Z} \setminus |\gamma q_2| < r\},$$
$$E_2 = \{(\alpha + m\beta)q_0 \setminus |\alpha + m\beta| < 2r(|s_1^0| + |s_2^0|) , \ \alpha, \beta \in \mathbb{Z}\},$$

and for $j = 1, 2$

$$E_{(j,3)} = \{(\alpha + m\beta)z_1 s_j^0 + \gamma t_j^2 + u \setminus (\alpha + m\beta)q_0 \in E_2 ,$$
$$(\alpha + m\beta)z_1 s_j^0 + \gamma t_j^2 + u| < r, \gamma \in E_1 , \ u \in \mathbb{Z}\},$$

are finite where $T^2 = (t_1^2, t_2^2)$. Then $G.Q_1 \cap K$ is finite; this implies that

$$\inf\{d(Q, Q_1) = \|\overrightarrow{QQ_1}\| \setminus Q \in G.Q_1\} = \alpha > 0.$$

Let

$$\varepsilon' = \min\left(\frac{1}{q_0}, \frac{|D_2|}{4q_0(|s_1^0| + |s_2^0|)}, \frac{\alpha}{2}\right),$$

and let U_1 (resp., U_2) be the open ball with center Q_1 (resp., Q_2) and with radius ε'.

We have

$$G.U_1 \cap U_2 = \emptyset.$$

Case 2: $S^0 = 0$.

We may prove this assertion for this case exactly as for the first one. Notice that in this case the group Γ_2^{02012} is a free abelian group on three generators.

2. The fact that the vectors $(T^0, 0)$, $(T^1, 0)$ and (T^2, q_2) are linearly independent permits us to see that the quotient M_2^{02012} is a compact manifold.

   The orientability of this manifold follows from each element of the group $Hp(2, 1; \mathbb{R})$ having positive determinant.

3. Following references [3] and [4], two compact connected complete locally affine manifolds M_1 and M_2 are homeomorphic if and only if the fundamental groups $\pi_1(M_1)$ and $\pi_1(M_2)$ are isomorphic; this implies that M_2^{02012} is homeomorphic to the torus T^3 if and only if the fundamental group $\pi_1(M_2^{02012}) = \Gamma_2^{02012}$ of M is an abelian group if and only if $S^0 = 0$. ■

Chapter 8

HOMOGENEOUS k-SYMPLECTIC G-SPACES

8.1 k-symplectic G-spaces

8.1.1   Strongly Hamiltonian systems

Let M be an $n(k+1)$-dimensional manifold equipped with a k-symplectic structure $(\theta^1, \ldots, \theta^k; E)$. Let $\mathfrak{F}$ be the foliation defined by the sub-bundle E. By the relationship

$$L_X \theta = i(X)d\theta + di(X)\theta$$

for every $\omega \in \Lambda^p(M)$ we see that the following properties are equivalent :

1. $L_X \theta^1 = \ldots = L_X \theta^k = 0$;

2. The Pfaffian forms $i(X)\theta^1, \ldots, i(X)\theta^k$ are closed;

where L_X is the Lie derivative with respect to X. Then a necessary and sufficient condition for an infinitesimal automorphism X of $\mathfrak{F}$ to be a Hamiltonian system is that

$$i(X)\theta^1, \ldots, i(X)\theta^k$$

are closed Pfaffian forms.

Recall that a strongly Hamiltonian system on M is a Hamiltonian system X such that there exists a smooth mapping $H : M \longrightarrow \mathbb{R}^k$ satisfying

$$j(X) = -dH$$

where $j(X) = (j^1(X), \ldots, j^k(X)) = (i(X)\theta^1, \ldots, i(X)\theta^k)$.

Let $\Xi(M)$ be the set of Hamiltonian systems and let $\mathfrak{S}(M)$ be the set of strongly Hamiltonian systems of the k-symplectic structure. The manifold M is assumed to be connected, then we have:

Proposition 8.1 *Let* H *be a Hamiltonian mapping and let* X_H *be the Hamiltonian system associated with* H. *The following properties are equivalent*

1. $j(X_H) = 0$;

2. $X_H = 0$;

3. H *is constant on* M.

Consequently the sequence of Lie algebras

$$0 \longrightarrow \mathbb{R}^k \xrightarrow{\text{inj}} \mathfrak{H}(M) \xrightarrow{\nu} \mathfrak{S}(M) \longrightarrow 0$$

is exact, where inj is the canonical injection and ν is the mapping from $\mathfrak{H}(M)$ into $\mathfrak{S}(M)$ defined by

$$\nu(H) = X_H$$

for each $H \in \mathfrak{H}(M)$.

Proposition 8.2 *The following properties are satisfied:*

1. $[\Xi(M), \Xi(M)] \subseteq \mathfrak{S}(M)$;

2. $\mathfrak{S}(M)$ *is an ideal of* $\Xi(M)$.

Proof. We have

$$i([X,Y])\theta^p = [L_X, i(Y)]\theta^p = L_X(i(Y)\theta^p) = -d(\theta^p(X,Y))$$

for all X and Y belonging to $\Xi(M)$, thus $[X,Y] \in \mathfrak{S}(M)$. The second property is immediate. ∎

8.1.2 k-symplectic actions

Let G be a Lie group, $\mathcal{G}$ the Lie algebra of G, and M be a smooth manifold of dimension $n(k+1)$ equipped with a k-symplectic structure $(\theta^1, \ldots, \theta^k; E)$, and let

$$\varphi : G \times M \longrightarrow M$$

be a smooth left action of the group G on M. For every $g \in G$ we denote by φ_g the diffeomorphism

$$x \longmapsto \varphi(g, x)$$

of M.

Definition 8.1 *We say that M is a k-symplectic G-space if for every $g \in G$ and $x \in M$ we have*

1. *$\varphi_g^* \theta^p = \theta^p$. for each p $(p = 1, \ldots, k)$;*

2. *$(\varphi_g)_* E_x = E_{\varphi(g,x)}$.*

If, in addition this action is transitive we say that M is a homogeneous k-symplectic G-space.

For every $X \in \mathcal{G}$ we denote by X_M the fundamental vector field associated with X with respect to this action. It is defined by

$$X_M(x) = \frac{d}{dt} \left(\varphi(\exp(tX), x) \right)_{\mid t=0}.$$

The vector field X_M is complete and the association $\sigma : X \longmapsto X_M$ from $\mathcal{G}$ into $\mathfrak{X}(M)$ is an injective anti-homomorphism of Lie algebras, that is,

$$[X, Y]_M = -[X_M, Y_M],$$

and it is well known, that if, in addition, the group G acts effectively on M then σ is injective.

Let $x \in M$; denote by $G.x$ the orbit of x under G, and let G_x be the isotropy subgroup of x in G; with respect to this action G_x is a closed subgroup of G and the homogeneous space G/G_x admits a natural structure of smooth manifold, the G-orbit $G.x$ will be equipped with its structure of smooth manifold in order that the bijection

$$g.G_x \longmapsto \varphi(g, x),$$

from G/G_x onto $G.x$ is an isomorphism of smooth manifolds.

For M not necessarily connected we have ([34]):

Proposition 8.3 *For every* $x \in M$ *and* y *belonging to the G-orbit* $G.x$ *the tangent space* $T_y(G.x)$ *of* $G.x$ *at the point* y *is the set of tangent vectors of type* $X_M(y)$ *such that* $X \in \mathcal{G}$ *and the Lie algebra* $\mathcal{G}_x$ *of* G_x *is formed by elements* $X \in \mathcal{G}$ *such that* $X_M(x) = 0$:

$$T_y(G.x) = \{X_M(y) \mid X \in \mathcal{G}\}$$

and

$$\mathcal{G}_x = \{X \in \mathcal{G} \mid X_M(x) = 0\}.$$

There is a natural action of the Lie group G on the real space $\hom(\mathcal{G}, \mathbb{R}^k)$, called the coadjoint representation and inducing the ordinary case for $k = 1$; this action is given by

$$\langle Ad_g^*(f), X \rangle = \langle f, Ad_{g^{-1}}(X) \rangle$$

for all $g \in G$, $f \in \hom(\mathcal{G}, \mathbb{R}^k)$ and $X \in \mathcal{G}$. The orbits of this action are called coadjoint G-orbits.

It follows immediately from the definition of a k-symplectic G-space that we have :

Proposition 8.4 *Let* M *be a k-symplectic G-space. For every* $X \in \mathcal{G}$ *the vector field* X_M *is a Hamiltonian system.*

Definition 8.2 *A k-symplectic G-space* M *is said to be strongly k-symplectic if for every element* $X \in \mathcal{G}$ *the vector field* X_M *is a strongly Hamiltonian system.*

Proposition 8.5 *For a k-symplectic G-space* M *to be strongly k-symplectic, it is necessary and sufficient that there exists a basis* $(X^1, \ldots, X^r)$ *of* $\mathcal{G}$ *whose the associated fundamental vector fields* $X_M^1, \ldots, X_M^r$ *are strongly Hamiltonian systems.*

Proposition 8.6 *A k-symplectic G-space* M *is strongly k-symplectic in the following cases:*

1. *The de Rham cohomology group* $H^1(M, \mathbb{R})$ *is trivial;*

2. $\mathcal{G}$ *verifies* $[\mathcal{G}, \mathcal{G}] = \mathcal{G}$.

Proof. Let M be a k-symplectic G-space and let $X \in \mathcal{G}$. Assume that the de Rham cohomology group $H^1(M, \mathbb{R})$ is trivial, then the Pfaffian forms $i(X)\theta^1, \ldots, i(X)\theta^k$ are exact. Therefore there exists $H \in C^\infty(M, \mathbb{R}^k)$ such that $j(X_H) = -dH$, thus M is a strongly k-symplectic G-space. If $[\mathcal{G}, \mathcal{G}] = \mathcal{G}$ then there exist $X^1, Y^1 \ldots, X^q, Y^q \in \mathcal{G}$, such that

$$X = \sum_{i=1}^{q} [X^i, Y^i]$$

consequently

$$X_M = -\sum_{i=1}^{\nu} [X_M^i, Y_M^i].$$

We have $[\Xi(M), \Xi(M)] \subseteq \mathfrak{S}(M)$ and $X_M \in \Xi(M)$ for $X \in \mathcal{G}$. Then

$$[X_M^i, Y_M^i] \in \mathfrak{S}(M)$$

for all i $(i = 1, \ldots, q)$, thus X_M is a strongly Hamiltonian system which proves that the k-symplectic G-space M is strongly k-symplectic. ∎

Definition 8.3 *Let M be a strongly k-symplectic G-space. A lift of M is a Lie algebras homomorphism J from $\mathcal{G}$ into $\mathfrak{H}(M)$ such that for each $X \in \mathcal{G}$, the vector field X_M coincides with the Hamiltonian system which is associated with the Hamiltonian mapping $-J(X)$; in other words, a homomorphism of Lie algebras J from $\mathcal{G}$ into $\mathfrak{H}(M)$ is a lift of M if the condition*

$$X_M = -X_{J(X)}$$

is satisfied for every $X \in \mathcal{G}$.

Definition 8.4 *A strongly k-symplectic G-space equipped with a lift is called a Hamiltonian G-space.*

8.2 Momentum mappings

The concept of momentum mapping was introduced in symplectic geometry by Smale and Souriau. In this section, we will generalize to the $k-$ symplectic case this notion. For a general presentation of these notions, see for example, [LBN−MRL].

Definition 8.5 *Let M be a Hamiltonian G-space equipped with a lift J. The momentum mapping of J is the mapping $\tilde{J}$ from M into $\hom(\mathcal{G}, \mathbb{R}^k)$ defined by*

$$\tilde{J}(x)(X) = J(X)(x)$$

for all $x \in M$ and $X \in \mathcal{G}$.

8.2.1   The canonical k-symplectic case

Consider the real space $\mathbb{R}^{n(k+1)}$ equipped with its canonical k-symplectic structure $(\theta^1, \ldots, \theta^k; E)$ defined by the forms

$$\theta^p = \sum_{i=1}^{n} dx^{pi} \wedge dx^i.$$

and by the sub-bundle E of $T\mathbb{R}^{n(k+1)}$ defined by the equations

$$dx^1 = \ldots = dx^n = 0,$$

where $(x^{pi}, x^i)_{1 \le p \le k, 1 \le i \le n}$ is the Cartesian coordinate system of $\mathbb{R}^{n(k+1)}$. Let M be the space $M = \mathbb{R}^{nk} \times (\mathbb{R}^n - \{0\})$ equipped with the k-symplectic structure induced by the canonical k-symplectic structure of $\mathbb{R}^{n(k+1)}$. The natural action

$$(S, X) \longmapsto SX$$

of the k-symplectic group $Sp(k, n; \mathbb{R})$ on M is transitive and effective, and M is a strongly homogeneous k-symplectic $Sp(k, n; \mathbb{R})$-space. In fact, every element X of $\mathfrak{sp}(k, n; \mathbb{R})$ can be written

$$X = \begin{pmatrix} A & \cdots & 0 & S^1 \\ & \ddots & & \vdots \\ 0 & & A & S^k \\ 0 & \cdots & 0 & -{}^t A \end{pmatrix},$$

where $A, S^1, \cdots, S^k$ are $n \times n$ matrices with real coefficients such that

$${}^t S^p = S^p$$

for each p $(p = 1, \ldots, k)$. The fundamental vector field X_M associated with X by the natural action of $Sp(k, n; \mathbb{R})$ on M is given by

$$X_M(x) = X.x$$

for all $x \in M = \mathbb{R}^{nk} \times (\mathbb{R}^n - \{0\})$. For $x = (x^{pi}, x^i)_{1 \leq p \leq k, 1 \leq i \leq n} \in M$, $A = (u_{ij})$ and $S^p = \left(s_{ij}^p \right)$ we have

$$X_M(x) = \sum_{i=1}^{n} \left(\sum_{j=1}^{n} \sum_{p=1}^{k} ((u_{ij} x^{pj} + s_{ij}^p x^j) \frac{\partial}{\partial x^{pi}} - u_{ji} x^j \frac{\partial}{\partial x^i}) \right)$$

thus

$$X_M(x) = \sum_{i=1}^{n} \left(-\sum_{p=1}^{k} \left(\sum_{j=1}^{n} \frac{\partial f_j}{\partial x^i} (x^1, \ldots, x^n) x^{pi} + \frac{\partial g^p}{\partial x^i} (x^1, \ldots, x^n) \right) \frac{\partial}{\partial x^{pi}} \right.$$
$$\left. + f_i(x^1, \ldots, x^n) \frac{\partial}{\partial x^i} \right)$$

where

$$f_j = -\sum_{i=1}^{n} u_{ji} x^i, \qquad g^p = -\frac{1}{2} \sum_{i,j=1}^{n} s_{ij}^p x^i x^j.$$

It results immediately that the mapping $H : M \longrightarrow \mathbb{R}^k$ whose components are

$$H^p = \sum_{j=1}^{n} f_j(x^1, \ldots, x^n) x^{pj} + g^p(x^1, \ldots, x^n)$$

is a Hamiltonian mapping of X_M. The association $X \longmapsto -H$ allows us to define, thanks to the preceding construction, a mapping

$$J : \mathfrak{sp}(k, n; \mathbb{R}) \longrightarrow \mathfrak{H}(M)$$

satisfying

$$J(X)(0) = 0.$$

and

$$X_M = -X_{J(X)}.$$

One verifies easily that J is a morphism of Lie algebras. Thus the mapping J confers upon M a structure of a Hamiltonian $Sp(k, n; \mathbb{R})$-space.

The components $\tilde{J}^p$ of the momentum mapping associated with the lift J is the mapping from M into $\mathfrak{sp}(k, n; \mathbb{R})^*$ defined by

$$\tilde{J}^p(x)(X) = -\left(\sum_{i,j=1}^n u_{ji} x^j x^{pi} + \frac{1}{2} \sum_{i,j=1}^n s_{ij}^p x^i x^j \right)$$

for all $x = (x^{pi}, x^i)_{1 \leq p \leq k, 1 \leq i \leq n}$ belonging to M and $X \in \mathfrak{sp}(k, n; \mathbb{R})$.

8.2.2   Some properties of the momentum mapping

Let M be a k–symplectic G-space. For all $H \in \mathfrak{H}(M)$ and $g \in G$, $\varphi_g^* H \in \mathfrak{H}(M)$; more precisely, we have:

$$X_{\varphi_g^* H} = (\varphi_g^{-1})_* X_H .$$

Let, now, M be a Hamiltonian G-space equipped with a lift J, then for all $X \in \mathcal{G}$ and $g \in G$ we have

$$X_{\varphi_g^* J(X)} = (\varphi_g^{-1})_* X_{J(X)} .$$

Thus follows:

Proposition 8.7 *Let M be a k-symplectic G-space. For every $g \in G$ the mapping φ_g^* is an isomorphism of Lie algebras from $\mathfrak{H}(M)$ onto itself.*

It is well known that for each $X \in \mathcal{G}$, $g \in G$ and $t \in \mathbb{R}$ we have

$$\exp(t Ad_g X) = g \exp(tX) g^{-1}.$$

By derivation with respect to t we obtain

$$(\varphi_g)_* X_M = (Ad_g X)_M .$$

If M is a Hamiltonian G-space with left J then we have

$$X_{\varphi_{g^{-1}}^* J(X)} = (Ad_g X)_M ,$$

for every $X \in \mathcal{G}$ and $g \in G$. In other words, for each $g \in G$ and $X \in \mathcal{G}$ the mapping

$$\Theta(g)(X) = (\varphi_{g^{-1}})^* J(X) - J(Ad_g X),$$

defined on M and with values in $\mathbb{R}^k$, is constant on M; hence there is a smooth mapping from G into $\hom(\mathcal{G}, \mathbb{R}^k)$, also denoted by Θ, such that for all $g \in G$ and $x \in M$ we have

$$\Theta(g) = \tilde{J}(\varphi_{g^{-1}}(x)) - Ad^*_{g^{-1}}(\tilde{J}(x)).$$

Let $c : \mathcal{G} \longrightarrow \hom(\mathcal{G}, \mathbb{R}^k)$ be the mapping defined by

$$c(g) = \Theta(g^{-1}).$$

for all $g \in G$. We have:

$$
\begin{aligned}
c(g_1 g_2) &= \tilde{J}(\varphi_{g_1}(\varphi_{g_2}(x))) - Ad^*_{g_1}(Ad^*_{g_2}(\tilde{J}(x))) \\
&= \tilde{J}(\varphi_{g_1}(\varphi_{g_2}(x))) - Ad^*_{g_1}(\tilde{J}(\varphi_{g_2}(x))) \\
&\quad + Ad^*_{g_1}(\tilde{J}(\varphi_{g_2}(x))) - Ad^*_{g_1}(Ad^*_{g_2}(\tilde{J}(x))) \\
&= c(g_1) + Ad^*_{g_1}(c(g_2))
\end{aligned}
$$

for all $g_1, g_2 \in G$ and $x \in M$, then the components c^p of c (which are mappings from G into $\mathcal{G}^*$) satisfy the condition

$$c^p(g_1 g_2) = c^p(g_1) + Ad^*_{g_1}(c^p(g_2))$$

characterizing the 1−cocycles of G with respect to the coadjoint representation, thus

$$c^T_g = Ad^*_g \circ c^T_e \circ L^T_{g^{-1}}$$

for all $g \in G$, where c^T denotes the tangent mapping of c, e the unit element of G and L the left-translation of G.

Proposition 8.8 *For all $g \in G$ and $X, Y \in \mathcal{G}$ we have:*

$$\Theta(g)([X, Y]) = 0.$$

Proof. In fact, we have

$$J(Ad_g[X, Y]) = J([Ad_g X, Ad_g Y]) = \{J(Ad_g X), J(Ad_g Y)\},$$

As $\Theta(g)(X)$ and $\Theta(g)(Y)$ are constant on M, thus

$$
\begin{aligned}
J(Ad_g[X, Y]) &= \{(\varphi_{g^{-1}})^* J(X), (\varphi_{g^{-1}})^* J(Y)\} \\
&= \varphi_{g^{-1}})^* \{J(X), J(Y)\} \\
&= (\varphi_{g^{-1}})^* J([X, Y]),
\end{aligned}
$$

yielding $\Theta(g)([X, Y]) = 0$. ∎

Proposition 8.9 *For all $g \in G$ and $X \in \mathcal{G}$, the mapping Θ is constant on the curve $\gamma(t) = g \exp(tX)$.*

Proof. For all $Y \in \mathcal{G}$, we have:

$$
\begin{aligned}
(\tfrac{d}{dt}\Theta(\gamma(t)_{|t=0})(Y) \ &= \ \tfrac{d}{dt}((\varphi_{g \exp(tX)^{-1}})^* J(Y))_{|t=0} - J(Ad_g[X,Y]) \\
&= dJ(Y) \circ (\tfrac{d}{dt}((\varphi_{\exp(-tX)} \circ \varphi_{g^{-1}})_{|t=0}) - J(Ad_g[X,Y]) \\
&= -dJ(Y)(X_M) \circ \varphi_g^{-1} - J(Ad_g[X,Y]) \\
&= (\varphi_g^*)^{-1}\{J(X), J(Y)\} - J(Ad_g[X,Y]) \\
&= (\varphi_g^*)^{-1} J([X,Y]) - J(Ad_g[X,Y]) \\
&= \Theta(g)([X,Y]) = 0.
\end{aligned}
$$

As Y is arbitrary

$$
(\frac{d}{dt}\Theta(\gamma(t)_{|t=0}) = 0.
$$

This proves the proposition. ∎

The previous proposition implies that $\Theta_e^T = 0$ and $c_e^T = 0$; consequently we have $c_g^T = 0$ for any $g \in G$. But Θ is constant on the unit component of G; thus we have

$$
\Theta(g)(X) = \Theta(e)(X) = J(X) - J(X) = 0,
$$

for all $g \in G$ and $X \in \mathcal{G}$, thus

$$
\varphi_g^* \circ J = J \circ Ad_g
$$

for every $g \in G$. We have proved the following proposition:

Proposition 8.10 *The momentum mapping $\tilde{J}$ is equivariant; that is, for every $g \in G$ we have*

$$
\tilde{J} \circ \varphi_g = Ad_g^* \circ \tilde{J}.
$$

Corollary 8.1 *Let M be a Hamiltonian G-space, let H be a G-invariant Hamiltonian mapping, that is, $H(\varphi(g,x)) = H(x)$ for all $g \in G$ and $x \in M$. Then the associated momentum mapping $\tilde{J}$ is constant on the integral curves of the vector field X_H.*

Consider now a strongly k-symplectic G-space, let $X^1, \ldots, X^r$ be a basis of $\mathcal{G}$, and let $H^1, \ldots, H^r$ be Hamiltonian mappings associated to $X_M^1, \ldots,$ X_M^r respectively. The linear mapping

$$J_0 : \mathcal{G} \longrightarrow \mathfrak{H}(M)$$

defined by

$$J_0(X^i) = -H^i$$

for each i $(i = 1, \ldots, r)$ satisfies the relation

$$X_{J_0(X)} = -X_M$$

for every $X \in \mathcal{G}$. If J_0 is a morphism of Lie algebras, then M is a Hamiltonian G-space. Suppose that J_0 is not a morphism of Lie algebras. Let

$$\mu : \mathcal{G} \times \mathcal{G} \longrightarrow \mathfrak{H}(M)$$

be the mapping defined by

$$\mu(X, Y) = \{J_0(X), J_0(Y)\} - J_0([X, Y]).$$

For all $(X, Y) \in \mathcal{G} \times \mathcal{G}$ we have

$$X_{\mu(X,Y)} = 0,$$

thus $\mu(X, Y)$ is constant on M, and consequently μ defines a skew-symmetric $\mathbb{R}$-bilinear mapping β from $\mathcal{G} \times \mathcal{G}$ into $\mathbb{R}^k$.

Proposition 8.11 *Under the previous hypothesis and notations we have the following properties*

1. *$\delta\beta = 0$;*

2. *If there exists $\alpha \in \hom(\mathcal{G}, \mathbb{R}^k)$ such that $\beta = \delta\alpha$, in particular if the de Rham cohomology group $H^2(M, \mathbb{R})$ is trivial, then M admits a lift (which is not unique in the general case);*

3. *If $H^2(M, \mathbb{R})$ is trivial and $[\mathcal{G}, \mathcal{G}] = \mathcal{G}$ the G-space M has an unique lift;*

4. *If β is not a coboundary, that is, β is not of the form $\delta\alpha$ where $\alpha \in \hom(\mathcal{G}, \mathbb{R}^k)$, then there exists a connected and simply connected Lie group $\widetilde{G(\beta)}$ with Lie algebra $\widetilde{\mathcal{G}(\beta)} = \mathcal{G} \times \mathbb{R}^k$ with respect to the law*

$$B((X, T), (Y, S)) = ([X, Y], \beta(X, Y)),$$

such that M is a Hamiltonian $\widetilde{G(\beta)}$-space.

Proof. Let $Z^1, Z^2, Z^3 \in \mathcal{G}$. We have:

$$\delta\beta(Z^1, Z^2, Z^3) = -\beta([Z^1, Z^2], Z^3) + \beta([Z^1, Z^3], Z^2)$$
$$-\beta([Z^2, Z^3], Z^1)$$
$$= -\{J_0([Z^1, Z^2]), J_0(Z^3)\} + \{J_0([Z^1, Z^3]), J_0(Z^2)\}$$
$$-\{J_0([Z^2, Z^3]), J_0(Z^1)\},$$

and for each p $(p = 1, \ldots, k)$ we have:

$$\{J_0([Z^1, Z^2]), J_0(Z^3)\}^p = -\theta^p([Z^1, Z^2]_M, Z_M^3)$$
$$= (i(Z_M^3)\theta^p)([Z^1, Z^2]_M)$$
$$= -[Z^1, Z^2]_M(J_0^p(Z^3))$$
$$= [Z_M^1, Z_M^2](J_0^p(Z^3))$$
$$= [X_{J_0(Z^1)}, X_{J_0(Z^2)}](J_0^p(Z^3))$$
$$= X_{\{J_0(Z^1), J_0(Z^2)\}}(J_0^p(Z^3))$$
$$= -(i(Z_M^3)\theta^p)(X_{\{J_0(Z^1), J_0(Z^2)\}})$$
$$= -\theta^p(X_{\{J_0(Z^1), J_0(Z^2)\}}, X_{J_0(Z^3)})$$
$$= -\{\{J_0(Z^1), J_0(Z^2)\}, J_0(Z^3)\}^p$$

then,

$$\{J_0([Z^1, Z^2]), J_0(Z^3)\} = -\{\{J_0(Z^1), J_0(Z^2)\}, J_0(Z^3)\}.$$

Consequently we have:

$$\delta\beta((Z^1, Z^2, Z^3) = \{\{J_0(Z^1), J_0(Z^2)\}, J_0(Z^3)\}$$
$$-\{\{J_0(Z^1), J_0(Z^3)\}, J_0(Z^2)\} + \{\{J_0(Z^2), J_0(Z^3)\}, J_0(Z^1)\}$$
$$= 0,$$

and the first assertion is proved.

Assume that there exists $\alpha \in \hom(\mathcal{G}, \mathbb{R}^k)$ such that $\beta = \delta\alpha$. Let J be the mapping from $\mathcal{G}$ into $\mathfrak{H}(M)$ defined by

$$J(X)(x) = J_0(X)(x) - \alpha(X)$$

for every $X \in \mathcal{G}$ and $x \in M$. We have:

$$(\{J(X), J(Y)\} - J[X, Y])(x) = \{J_0(X), J_0(Y)\}(x) - J_0[X, Y](x)$$
$$+\alpha([X, Y]) = \mu(X, Y)(x) + \alpha([X, Y])$$
$$= \beta(X, Y) + \alpha([X, Y])$$
$$= \delta\alpha(X, Y)(x) + \alpha([X, Y]) = 0,$$

for all $X, Y \in \mathcal{G}$ and $x \in M$. The mapping J defines a lift of M, which proves the second assertion.

We now prove the uniqueness of the lift. Suppose that $H^2(M, \mathbb{R}) = \{0\}$ and $[\mathcal{G}, \mathcal{G}] = \mathcal{G}$. The second property proves that the k-symplectic G-space admits a lift. Let J_1, J_2 be two lifts of M. The relationship

$$X_{J_1(X)} = X_{J_2(X)},$$

for any $X \in \mathcal{G}$, proves that the mapping

$$J_1(X) - J_2(X),$$

from M into $\mathbb{R}^k$ is constant for every $X \in \mathcal{G}$; then $J_1 - J_2$ defines a mapping from $\mathcal{G}$ into $\mathbb{R}^k$ which is a homomorphism of Lie algebras from $\mathcal{G}$ into $\mathbb{R}^k$; therefore

$$(J_1 - J_2)([X, Y]) = 0$$

for all $X, Y \in \mathcal{G}$. Thus, the restriction of the mapping $J_1 - J_2$ to $[\mathcal{G}, \mathcal{G}]$ is zero, but $[\mathcal{G}, \mathcal{G}] = \mathcal{G}$, therefore $J_1 - J_2 = 0$ and the assertion 3 is proved.

Now let us prove the property 4. Suppose that β is not a coboundary. Let $\widetilde{\mathcal{G}(\beta)}$ be the product space

$$\widetilde{\mathcal{G}(\beta)} = \mathcal{G} \times \mathbb{R}^k$$

and B the mapping

$$B \; : \; \widetilde{\mathcal{G}(\beta)} \times \widetilde{\mathcal{G}(\beta)} \; \longrightarrow \; \widetilde{\mathcal{G}(\beta)}$$
$$(u, v) \; \longmapsto \; B(u, v) = ([X, Y], \beta(X, Y))$$

where $u = (X, T)$ and $v = (Y, S) \in \widetilde{\mathcal{G}(\beta)}$. This mapping is a $\mathbb{R}$-bilinear skew-symmetric mapping. The relationship $\delta\beta = 0$ proves that the Jacobi identity is satisfied, thus $(\widetilde{\mathcal{G}(\beta)}, B)$ is a real Lie algebra. Let $J : \widetilde{\mathcal{G}(\beta)} \longrightarrow \mathfrak{H}(M)$ be the mapping given by

$$J(u) = J_0(X) + T$$

for all $u = (X, T) \in \widetilde{\mathcal{G}(\beta)}$. We have

$$\{J(u), J(v)\} - J(B(u, v)) = \{J_0(X), J_0(Y)\} - J_0([X, Y]) - \beta(X, Y) = 0$$

for all $u = (X, T)$, $v = (Y, S) \in \widetilde{\mathcal{G}(\beta)}$, thus J is a morphism of Lie algebras from $\widetilde{\mathcal{G}(\beta)}$ into $\mathfrak{H}(M)$ satisfying

$$X_M = -X_{J(X)}$$

for all $u = (X, T) \in \widetilde{\mathcal{G}(\beta)}$. Let $\widetilde{G(\beta)}$ be a connected and simply connected Lie group with Lie algebra $\widetilde{\mathcal{G}(\beta)}$ and let $\widetilde{H(\beta)}$ be the connected subgroup of $\widetilde{G(\beta)}$ having as its Lie algebra $\widetilde{\mathcal{H}(\beta)} = 0_{\mathcal{G}} \times \mathbb{R}^k$. The Lie subalgebra $\widetilde{\mathcal{H}(\beta)}$ is an ideal of $\widetilde{\mathcal{G}(\beta)}$, thus $\widetilde{\mathcal{H}(\beta)}$ is a normal subgroup of $\widetilde{G(\beta)}$. The quotient space

$$\widetilde{Q(\beta)} = \frac{\widetilde{G(\beta)}}{\widetilde{H(\beta)}}$$

is a connected and simply connected Lie group whose Lie algebra is isomorphic to $\mathcal{G}$; thus there exists a smooth covering

$$p : \widetilde{Q(\beta)} \longrightarrow G$$

with base space G and with total space $\widetilde{Q(\beta)}$. Let $\tilde{p}$ be the mapping from $\widetilde{G(\beta)}$ into G defined by

$$\tilde{p}(\tilde{g}) = p(\tilde{g}\widetilde{H(\beta)})$$

for every $\tilde{g} \in \widetilde{G(\beta)}$. There is a smooth action of the group $\widetilde{G(\beta)}$ on M defined by

$$\tilde{\varphi}(\tilde{g}, x) = \varphi(\tilde{p}(\tilde{g}), x)$$

for each $\tilde{g} \in \widetilde{G(\beta)}$ and $x \in M$. With respect to this action we have

$$X_{J(u)} = -u_M = -X_M$$

for all $u = (X, T) \in \widetilde{\mathcal{G}(\beta)}$, that is, J is a lift of the $\widetilde{G(\beta)}$-space M; this proves that the manifold M is a Hamiltonian $\widetilde{G(\beta)}$-space. ∎

8.3   The Heisenberg group of rank k

Consider the space $M = \mathbb{R}^{n(k+1)}$ equipped with its canonical k-symplectic structure defined by the 2-forms

$$\theta^p = \sum_{i=1}^{n} dx^{pi} \wedge dx^i$$

and by the sub-bundle E of $T\mathbb{R}^{n(k+1)}$ whose equations are $dx^1 = \dots = dx^n = 0$ and where $(x^{pi}, x^i)_{1 \le p \le k, 1 \le i \le n}$ is the Cartesian coordinates system of $\mathbb{R}^{n(k+1)}$. The additive group $G = \mathbb{R}^{n(k+1)}$ acts transitively on M by the smooth mapping φ given by

$$\varphi(g, x) = x + g.$$

This action is effective (it is also free). The Lie algebra $\mathcal{G}$ of G is the space $\mathbb{R}^{n(k+1)}$ endowed with its trivial structure of the Lie algebra ($[X, Y] = 0$ for all X, Y). The adjoint and coadjoint actions of G on M and $\hom(\mathcal{G}, \mathbb{R}^k)$ are also trivial. For every $X \in \mathcal{G}$, the fundamental vector field X_M associated to X is defined by:

$$X_M(x) = X$$

for each $x \in M$.

Proposition 8.12 *The manifold M is a k-symplectic G-space.*

The family

$$\left(\frac{\partial}{\partial x^{pi}}, \frac{\partial}{\partial x^i} \right)_{1 \le p \le k, 1 \le i \le n}$$

defined by the derivatives associated to the Cartesian coordinates system $(x^{pi}, x^i)_{1 \le p \le k, 1 \le i \le n}$ form a basis of $\mathcal{G}$. The mappings H^{pi}, H^i from M into $\mathbb{R}^k$ defined by

$$H^{pi}(x) = -(\delta_1^p x^i, \dots, \delta_k^p x^i)$$

and

$$H^i(x) = -(x^{1i}, \dots, x^{ki})$$

satisfy the relations

$$X_{H^{pi}} = \left(\frac{\partial}{\partial x^{pi}} \right)_M$$

and

$$X_{H^i} = \left(\frac{\partial}{\partial x^i} \right)_M.$$

The $\mathbb{R}$-linear mapping $J_0 : \mathcal{G} \longrightarrow \mathfrak{H}(M)$ defined by

$$J_0 \left(\frac{\partial}{\partial x^{pj}} \right) = -H^{pj}$$

and

$$J_0 \left(\frac{\partial}{\partial x^j} \right) = -H^j.$$

satisfies the equality

$$X_M = -X_{J_0(X)}$$

for every $X \in \mathcal{G}$, then M is a strongly k-symplectic G-space. But $\mathcal{G}$ is an abelian Lie algebra in contrast to $\mathfrak{H}(M)$, therefore J_0 is not a homomorphism of Lie algebras and the mapping β defined above is not a coboundary. Consider the Lie algebra

$$\widetilde{\mathcal{G}(\beta)} = \mathcal{G} \times \mathbb{R}^k$$

equipped with the law

$$B((X,T),(Y,S)) = (0, \beta(X,Y)),$$

where

$$\beta(X,Y) = \{J_0(X), J_0(Y)\} - J_0([X,Y]) = \{J_0(X), J_0(Y)\}.$$

Let $\tilde{H}_{(n,k)}$ be the Lie group

$$\widetilde{H_{(n,k)}} = \mathbb{R}^{n(k+1)} \times \mathbb{R}^k$$

whose the group law is given by

$$u.v = (X + Y, T + S + \frac{1}{2}\beta(X,Y))$$

for all $u = (X,T), v = (Y,S) \in \tilde{H}_{(n,k)}$.

Definition 8.6 *The Heisenberg group of rank k is the connected and simply connected Lie group of dimension $(k+1)n+k$ whose derived group coincides with the center.*

This group has a left-invariant contact r-structure (see chapter 4). It plays the same role as the Heisenberg group with respect to contact structures. It can be represented as the group $H_{(n,k)}$ of matrices of the form

$$\begin{pmatrix} I_k & P & T \\ 0 & I_n & Q \\ 0 & 0 & 1 \end{pmatrix},$$

where I_n is the identity matrix of rank n. The Lie algebra $\mathcal{H}_{(n,k)}$ of the Heisenberg group of rank k is the Lie algebra formed by the matrices of the type

$$(X,Y,T) = \begin{pmatrix} 0 & X & T \\ 0 & 0 & Y \\ 0 & 0 & 0 \end{pmatrix},$$

where X is $k \times n$ real matrix and T (resp., Y) is a column matrix of length k (resp., n).

Proposition 8.13 $\widetilde{H_{(n,k)}}$ *is a connected and simply connected Lie group with a Lie algebra* $\mathcal{G}(\beta)$ *and it is isomorphic to the Heisenberg group* $H(n,k)$ *of rank* k.

Proof. Let

$$(X,Y,T) = \begin{pmatrix} 0 & X & T \\ 0 & 0 & Y \\ 0 & 0 & 0 \end{pmatrix}$$

be an element of $\mathcal{H}_{(n,k)}$. The mapping

$$(X,Y,T) \longmapsto ((X,Y),T)$$

from $\mathcal{H}_{(n,k)}$ into $\widetilde{\mathcal{G}(\beta)}$ is an isomorphism of Lie algebras. ∎

Proposition 8.14 *The mapping*

$$J : \mathcal{H}_{(n,k)} \longrightarrow \mathfrak{H}(M)$$

defined by

$$\forall (X,Y,T) \in \mathcal{H}_{(n,k)}, \ J(X,Y,T) = J_0(X,Y) + T,$$

is a lift of M, *therefore* M *is a Hamiltonian* $H_{(n,k)}$*-space.*

The corresponding momentum mapping is the mapping

$$\tilde{J} : M \longrightarrow \hom(\mathcal{H}_{(n,k)}, \mathbb{R}^k)$$

given by

$$\tilde{J}(x)(X,Y,T) = J_0(X,Y)(x) + T$$

for all $x \in M$ and $(X,Y,T) \in \mathcal{H}_{(n,k)}$.

8.4   Coadjoint G-orbits

Let M be a homogeneous Hamiltonian G-space equipped with a lift J with momentum mapping $\tilde{J}$ and let $\mathcal{H}^E$ be the set

$$\mathcal{H}^E = \{X \in \mathcal{G} \mid X_M \in \Gamma(E)\}.$$

The sub-bundle E of TM is integrable, thus $\mathcal{H}^E$ is a Lie subalgebra of $\mathcal{G}$.

Proposition 8.15 *For every* $x \in M$, *the tangent space of the leaf of* $\mathfrak{F}$ *passing through the point* x *is given by:*

$$E_x = \{X_M(x) \setminus X \in \mathcal{H}^E\}.$$

Proof. It results immediately from the definition of $\mathcal{H}^E$ that we have

$$X_M(x) \in E_x$$

for every $X \in \mathcal{H}^E$.

   Conversely, let $v_x \in E_x$. Considering the space M equipped with its structure of an nk-dimensional manifold whose components are the leaves of the foliation $\mathfrak{F}$. The tangent space of M at the point x with respect to this manifold structure is E_x. The action of G on M is transitive, thus there exists $X \in \mathcal{G}$ such that

$$v_x = X_M(x).$$

Hence for the initial structure of differential manifold of M, X_M is a cross-section of E, thus $X \in \mathcal{H}^E$; on the other hand,

$$v_x = X_M(x)$$

with $X \in \mathcal{H}^E$. ∎

Proposition 8.16 *If the action of* G *on* M *is effective, then* $\mathcal{H}^E$ *is an abelian Lie subalgebra of* $\mathcal{G}$.

Proof. The fundamental vector fields are Hamiltonian systems, then for all $X, Y \in \mathcal{H}^E$ and p $(p = 1, \ldots, k)$ we have

$$L_{X_M} \theta^p = L_{Y_M} \theta^p = 0,$$

yielding

$$
\begin{aligned}
i([X,Y]_M)\theta^p &= -i([X_M,Y_M])\theta^p = -[L_{X_M},i(Y_M)]\theta^p \\
&= -L_{X_M}(i(Y_M)\theta^p) = d(\theta^p(X_M,Y_M)) \\
&= 0
\end{aligned}
$$

for all $p\,(p=1,\ldots,k)$, because X_M and Y_M are tangent vector of the leaves of $\mathfrak{F}$, therefore

$$
[X,Y]_M \in C_x(\theta^1)\cap\ldots\cap C_x(\theta^k).
$$

As a consequence,

$$
[X,Y]_M = 0.
$$

This action of G on M is assumed to be effective, thus we have

$$
[X,Y] = 0.
$$

This proves that $\mathcal{H}^E$ is an abelian Lie algebra. ∎

Example 24 *Consider the case of the canonical k-symplectic space. In this case $\mathcal{H}^E$ is the Lie algebra $\mathfrak{tp}(k,n;\mathbb{R})$ of the Lie group $Tp(k,n;\mathbb{R})$, it is the subgroup of $Sp(k,n;\mathbb{R})$ spanned by the k-symplectic transvections. The Lie subalgebra $\mathfrak{tp}(k,n;\mathbb{R})$ is an abelian ideal of $\mathfrak{sp}(k,n;\mathbb{R})$*

For every $X \in \mathcal{G}$ we denote by X^* the fundamental vector field associated with X with respect to the coadjoint action of the group G on $\hom(\mathcal{G},\mathbb{R}^k)$; it results from the equality

$$
\frac{d}{dt}\Big|_{t=0} Ad_{\exp(tX)}Y = [X,Y],
$$

that we have:

Proposition 8.17 *For all $X,Y \in \mathcal{G}$ and $f \in \hom(\mathcal{G},\mathbb{R}^k)$ we have:*

$$
\langle Y^*(f),X\rangle = -\langle f,[X,Y]\rangle.
$$

Let O be a coadjoint G-orbit, let $f = (f^p)_{1\leq p\leq k}$ be a point belonging to O, and let $X_1,X_2 \in \mathcal{G}$ such that

$$
X_1^*(f) = X_2^*(f).
$$

For all p $(p = 1, \ldots, k)$ and $Y \in \mathcal{G}$ we have

$$\langle f^p, [X_1, Y] \rangle = \langle f^p, [X_1 - X_2, Y] \rangle + \langle f^p, [X_2, Y] \rangle .$$

In fact, $X_1 - X_2$ belongs to the Lie algebra $\mathcal{G}_f$ of the isotropy group G_f of f with respect to the coadjoint action, thus

$$\langle f^p, [X_1 - X_2, Y] \rangle = \langle (X_1 - X_2)^*(f^p), Y \rangle = 0,$$

consequently,

$$\langle f^p, [X_1, Y] \rangle = \langle f^p, [X_2, Y] \rangle .$$

Thus we have

Proposition 8.18 *Under the previous hypothesis and notations there exists on the coadjoint G-orbit O, closed differential forms $\Omega^1, \ldots, \Omega^k \in \Lambda^2(O)$ given by*

$$\Omega_f^p(X^*(f), Y^*(f)) = - \langle f^p, [X, Y] \rangle$$

for all $f = (f^p)_{1 \leq p \leq k}$ belonging to the coadjoint G-orbit O, $X^*(f), Y^*(f) \in T_f O$, and $p = 1, \ldots, k$.

Proof. Let $f = (f^p)_{1 \leq p \leq k} \in O$, $X, Y, Z \in \mathcal{G}$ and $p = 1, \ldots, k$. We have

$$\Omega_f^p([X^*, Y^*]\ \ , Z^*) - \Omega_f^p([X^*, Z^*], Y^*) + \Omega_f^p([Y^*, Z^*], X^*)$$
$$= \langle f^p, [[X, Y], Z] \rangle - \langle f^p, [[X, Z], Y] \rangle$$
$$+ \langle f^p, [[Y, Z], X] \rangle = 0.$$

Therefore,

$$d\Omega_f^p(X^*, Y^*, Z^*) = X_f^*(\Omega_f^p(Y^*, Z^*)) - Y_f^*(\Omega_f^p(X^*, Z^*))$$
$$+ Z_f^*(\Omega_f^p(X^*, Y^*)).$$

But,

$$X_f^*(\Omega_f^p(Y^*, Z^*)) = \tfrac{d}{dt}_{|t=0}(\psi(Ad^*_{\exp(tX)}(f)))$$
$$= \tfrac{d}{dt}_{|t=0}(\Omega^p_{Ad^*_{\exp(tX)}(f)}(Y^*, Z^*))$$
$$= - \tfrac{d}{dt}_{|t=0} \langle f^p \circ Ad_{\exp(-tX)}, [Y, Z] \rangle$$
$$= \langle f^p, [X, [Y, Z]] \rangle$$

where $\psi = \Omega^p(Y^*, Z^*)$; hence

$$d\Omega_f^p(X^*, Y^*, Z^*) = \langle f^p, [X, [Y, Z]] \rangle - \langle f^p, [Y, [X, Z]] \rangle$$
$$+ \langle f^p, [Z, [X, Y]] \rangle$$
$$= 0,$$

which proves that the differential forms Ω^p are closed. ■

Proposition 8.19 *If the Lie group G acts effectively on M then*

$$\Omega^p(X^*(f), Y^*(f)) = 0$$

for all $X, Y \in \mathcal{H}^E$, $f \in O$ and $p = 1, \ldots, k$.

Let M be a homogeneous Hamiltonian G-space equipped with a lift J, let $\tilde{J}$ be the momentum mapping associated to the lift J and let x_0 be a point of M. The Lie group G acts transitively on M, then for each x belonging to M, there exists $g \in G$ such that $x = \varphi(g, x_0)$; the equivariance of the momentum mapping $\tilde{J}$ implies that

$$\tilde{J}(x) = \tilde{J}(\varphi(g, x_0)) = Ad_g^*(\tilde{J}(x_0)),$$

which proves that $\tilde{J}(M)$ is a coadjoint G-orbit. Let $f \in \hom(\mathcal{G}, \mathbb{R}^k)$, O be a coadjoint G-orbit of f and let J_f be the mapping from $\mathcal{G}$ into $C^\infty(O, \mathbb{R}^k)$ defined by

$$J_f(X)(g) = \langle g, X \rangle$$

for all $X \in \mathcal{G}$ and $g \in O$. We have

$$\begin{aligned}
\tilde{J}^*(J_f(X))(x) &= J_f(X)(\tilde{J}(x)) \\
&= \left\langle \tilde{J}(x), X \right\rangle \\
&= J(X)(x)
\end{aligned}$$

for every $X \in \mathcal{G}$ and $x \in M$, hence

$$\tilde{J}^* \circ J_f = J.$$

It follows that we have

Lemma 25 *For every $X \in \mathcal{G}$ if $J_f(X)$ is constant on O then $J(X)$ is constant on M.*

The momentum mapping being equivariant, then for all $X \in \mathcal{G}$ and $x \in M$ we have:

$$\begin{aligned}
(\tilde{J}_* X_M)(\tilde{J}(x)) &= \tilde{J}_x^T X_M(x) = \tilde{J}_x^T \left(\tfrac{d}{dt}_{|t=0} \varphi(\exp(tX), x) \right) \\
&= \tfrac{d}{dt}_{|t=0} \tilde{J}(\varphi(\exp(tX), x)) \\
&= \tfrac{d}{dt}_{|t=0} Ad^*_{\exp(tX)}(\tilde{J}(x)) \\
&= X^*(\tilde{J}(x)),
\end{aligned}$$

thus,

$$\tilde{J}_* X_M = X^*.$$

Let $X, Y \in \mathcal{G}$, let $f \in \hom(\mathcal{G}, \mathbb{R}^k)$, let $O = G.f$ be the coadjoint G-orbit of f, let $g \in O$ and let $p = 1, \ldots, k$. We have

$$
\begin{aligned}
d(J_f^p(X))_g(Y^*) &= \tfrac{d}{dt}\big|_{t=0} J_f^p(X)\left(Ad^*_{\exp(tY)}(g)\right) \\
&= \tfrac{d}{dt}\big|_{t=0} \left\langle Ad^*_{\exp(tY)}(g^p), X \right\rangle \\
&= \tfrac{d}{dt}\big|_{t=0} \left\langle g^p, Ad_{\exp(-tY)}X \right\rangle \\
&= \left\langle g^p, [X, Y] \right\rangle \\
&= -\Omega_q^p(X^*(g), Y^*(g)) \\
&= -(i(X^*)\Omega_g^p)(Y^*).
\end{aligned}
$$

We have proved the next proposition:

Proposition 8.20 *For all* $X \in \mathcal{G}$ *and* $p = 1, \ldots, k$ *we have*

$$i(X^*)\Omega^p = -d(J_f^p(X)).$$

Let $X \in \mathcal{G}$, $f \in \hom(\mathcal{G}, \mathbb{R}^k)$ and $O = G.f$. The three preceding propositions show that the equality $\tilde{J}_* X_M = 0$ implies that $J_f(X)$ is constant on O; hence $J(X)$ is constant on M. Consequently $i(X_M)\theta^p = 0$ for every $p = 1, \ldots, k$. It results from the definition of a k-symplectic structure that $X_M = 0$ and that the mapping $\tilde{J}$ is an immersion; in particular, we have $\dim G \leq \dim (O)$. The equivariance of the momentum mapping proves that

$$G_{x_0} \subseteq G_{\tilde{J}(x_0)},$$

where G_{x_0}, $G_{\tilde{J}(x_0)}$ are the isotropy groups of x_0 and $\tilde{J}(x_0)$, respectively; thus $\dim(M) \geq \dim(O)$. We deduce that the identity components $G_{x_0}^e$ and $G_{\tilde{J}(x_0)}^e$ coincide; consequently the quotient space

$$\frac{G_{\tilde{J}(x_0)}}{G_{x_0}}$$

is discrete. This proves that

$$\tilde{J} : M = \frac{G}{G_{x_0}} \longrightarrow O = \frac{G}{G_{\tilde{J}(x_0)}}$$

is a fibration with discrete fibres, hence it is a covering space. We have shown the following result:

Proposition 8.21 *Let M be a homogeneous Hamiltonian G-space with a lift J and with associated momentum mapping $\tilde{J}$. Then $\tilde{J}(M)$ is a coadjoint G-orbit, and the mapping $\tilde{J} : M \longrightarrow O$ from M onto the coadjoint G-orbit O defines a covering space.*

One assumes that the Lie group G acts transitively and effectively on the manifold M. Let $E^{(O)}$ be the sub-bundle of TO defined by:

$$E_g^{(O)} = \{X_g^* \setminus X \in \mathcal{H}^E\}$$

for every $g \in O$. The previous proposition permits to see that we have:

1. The manifold O is of dimension $n(k+1)$;

2. The sub-bundle $E^{(O)}$ of TO is of codimension n and is integrable.

Proposition 8.22 *Let M be a homogeneous Hamiltonian G-space, let x_0 be a point of M, and let $f = \tilde{J}(x_0)$. Then*

$$(\Omega^1, \ldots, \Omega^k; E^{(O)})$$

defines a k-symplectic structure on $O = G.f$.

Proof. Let $g = (g^p)_{1 \leq p \leq k}$ be a point of O and let $X \in \mathcal{G}$. If we have $i(X^*)\theta^p = 0$ for every $p = 1, \ldots, k$, then

$$\langle g^p, [Y, X] \rangle = \langle X^*(g^p), Y \rangle = 0$$

for every $Y \in \mathcal{G}$; therefore, $X^*(g) = 0$, which proves the first condition of the k-symplectic structure, as for the second condition, it follows from the property

$$\Omega^p(X^*(f), Y^*(f)) = 0$$

for all $X, Y \in \mathcal{H}^E$. ∎

Proposition 8.23 *Under the hypothesis and notations of the previous proposition we have :*

1. *The coadjoint G-orbit $O = G.f$ of f is a homogeneous Hamiltonian G-space with a lift J_f;*

2. *The momentum mapping $\tilde{J}$ is a morphism of k-symplectic manifolds; more precisely we have:*

(a) $\tilde{J}\Omega^p = \theta^p$, for every $p = 1, ..., k$;

(b) $\tilde{J}_*(E) = E^{(O)}$.

Proof. For all $X, Y \in \mathcal{G}$, $p\,(p = 1, ..., k)$ and $h \in O$ we have:

$$(\psi_g^* \Omega^p)_h(X^*(h), Y^*(h)) = \Omega^p_{\psi_g(h)}\left((\psi_g)_h^T X^*(h), (\psi_g)_h^T Y^*(h)\right),$$

where $\psi_g = Ad_g^*$. But,

$$(\psi_g)_h^T X^*(h) = ((\psi_g)_* X^*)_h = (Ad_g X)^*(h),$$

therefore,

$$
\begin{aligned}
(\psi_g^* \Omega^p)_h(X^*(h), Y^*(h)) &= \Omega^p_{\psi_g(h)}((Ad_g X)^*(h), (Ad_g Y)^*(h)) \\
&= -\left\langle h^p, Ad_{g^{-1}}([Ad_g X, Ad_g Y])\right\rangle \\
&= -\left\langle h^p, [X, Y]\right\rangle \\
&= \Omega^p_h(X^*(h), Y^*(h)).
\end{aligned}
$$

One has proved the first condition of the definition of a strongly k-symplectic G-space; as for the second condition, it results from the relationship $\tilde{J}_* X_M = X^*$. The relation

$$i(X^*)\Omega^p = -d(J_f^p(X))$$

proves that the orbit O is a strongly k-symplectic G-space. Let $X, Y \in \mathcal{G}$, $p\,(p = 1, ..., k)$ and $h \in O$. We have:

$$
\begin{aligned}
\{J_f(X), J_f(Y)\}(h) &= -(\Omega^1_h(X_{J_f(X)}, X_{J_f(Y)}), ..., \Omega^k_h(X_{J_f(X)}, X_{J_f(Y)})) \\
&= -(\Omega^1_h(X^*(h), Y^*(h)), ..., \Omega^k_h((X^*(h), Y^*(h))) \\
&= \langle h, [X, Y]\rangle \\
&= J_f([X, Y])(h),
\end{aligned}
$$

then, J_f is a morphism of Lie algebras from $\mathcal{G}$ into $\mathfrak{H}(O)$, consequently the coadjoint G-orbit O is a homogeneous Hamiltonian G-space, which proves the property 1. The property (b) comes from the relationship $\tilde{J}_* X_M = X^*$ and the fact that $\tilde{J}(M)$ is a coadjoint G-orbit and from the mapping $\tilde{J}$ from M into O defining a covering space. Now let $X, Y \in \mathcal{G}$, $p\,(p = 1, ..., k)$

and $x_0 \in M$. We have:

$$
\begin{aligned}
(\tilde{J}^*\Omega^p)_{x_0}(X_M(x_0), Y_M(x_0)) &= \Omega^p_{\tilde{J}(x_0)}((\tilde{J}_*X_M)(x_0), (\tilde{J}_*Y_M)(x_0)) \\
&= \Omega^p_{\tilde{J}(x_0)}(X^*_M(\tilde{J}(x_0)), Y^*_M(\tilde{J}(x_0))) \\
&= -\left\langle \tilde{J}^p(x_0), [X, Y] \right\rangle \\
&= -\langle J^p([X, Y]), x_0 \rangle \\
&= \{J(X), J(Y)\}^p(x_0) \\
&= \theta^p_{x_0}(X_M(x_0), Y_M(x_0)) \ .
\end{aligned}
$$

This completes the proof. ∎

Chapter 9

GEOMETRIC PRE-QUANTIZATION

The aim of this chapter is a discussion about the geometric pre-quantization of the k−symplectic manifolds. In order to fix the terminology used, we begin by recalling some basic results of Cĕch cohomology, and complex bundle lines. For the proofs the reader is referred to Wallach [54] and Warner [55]. We advise the reader to use the first and second sections for reference purpose only.

9.1 The de Rham theorem

Let M be a paracompact connected smooth manifold. Recall that the p-th *de Rham cohomology group* of M is the quotient of the vector space of closed p-forms modulo the subspace of exact p-forms:

$$H_d^p(M, \mathbb{K}) = \frac{\ker \left(d : \Lambda^p(M, \mathbb{K}) \longrightarrow \Lambda^{p+1}(M, \mathbb{K}) \right)}{d\Lambda^{p-1}(M, \mathbb{K})},$$

where $\mathbb{K} = \mathbb{R}$ or $\mathbb{C}$, $\Lambda^p(M, \mathbb{K})$ is the set of $\mathbb{K}$-valued differential forms of degree p on M, and $p = 0, 1, 2, \ldots,$ $\Lambda^{-1}(M, \mathbb{K}) = \{0\}$.

The de Rham theorem relates $H_d^p(M, \mathbb{K})$ with $H^p(M, \mathbb{K})$, where $H^p(M, \mathbb{K})$ is one of the usual cohomology groups on M, in our case we use the Cĕch cohomology groups. We recall now the definition of the Cĕch cohomology groups. Let $\mathfrak{U} = (U_i)_{i \in I}$ be an open covering of M.

A collection $\left(U_{i_0}, U_{i_1}, \ldots, U_{i_p} \right)$ of members of the covering $\mathfrak{U}$ such that

$$U_{i_0, i_1, \ldots, i_p} = U_{i_0} \cap U_{i_1} \cap \ldots \cap U_{i_p} \neq \emptyset$$

is called a p-simplex. If $\sigma = (U_{i_0}, U_{i_1}, \ldots, U_{i_p})$ is a p-simplex, its support $|\sigma|$ is by definition

$$|\sigma| = U_{i_0} \cap U_{i_1} \cap \ldots \cap U_{i_p}.$$

Let G be an abelian Lie group written additively. A p-cochain on M with values in G is a rule c which associates with each p-simplex $(U_{i_0}, U_{i_1}, \ldots, U_{i_p})$ of $\mathfrak{U}$ an element

$$c_{i_0, i_1, \ldots, i_p} \in G.$$

Let $C^p(\mathfrak{U}, G)$ be the set of all p-cochains with values in G. We define a structure of an abelian group on $C^p(\mathfrak{U}, G)$ by taking

$$(c + f)_{i_0, i_1, \ldots, i_p} = c_{i_0, i_1, \ldots, i_p} + f_{i_0, i_1, \ldots, i_p}.$$

Let δ be the mapping from $C^p(\mathfrak{U}, G)$ into $C^{p+1}(\mathfrak{U}, G)$ defined by

$$(\delta c)_{i_0, i_1, \ldots, i_{p+1}} = \sum_{j=0}^{p+1} (-1)^j c_{i_0, i_1, \ldots, \widehat{i_j}, \ldots, i_p},$$

here the hat over an index means that this index is omitted. An immediate consequence of the definition is that

$$\delta^2 = \delta \circ \delta = 0.$$

Let $H^p(\mathfrak{U}, G)$ be the quotient group

$$H^p(\mathfrak{U}, G) = \frac{\ker \left(\delta : C^p(\mathfrak{U}, G) \longrightarrow C^{p+1}(\mathfrak{U}, G) \right)}{\delta C^{p-1}(\mathfrak{U}, G)},$$

where $p = 0, 1, 2, \ldots$, and $C^{-1}(\mathfrak{U}, G) = \{0\}$.

Let $\mathfrak{U}$ and $\mathfrak{V}$ be two open coverings of M such that $\mathfrak{U}$ refines $\mathfrak{V}$ $(\mathfrak{V} \prec \mathfrak{U})$, then there exists a mapping $\tau : \mathfrak{U} \longrightarrow \mathfrak{V}$ such that $\tau(\mathfrak{U}) \subseteq \mathfrak{U}$, and for each $c \in C^p(\mathfrak{U}, G)$ we define a mapping $\tau_{\mathfrak{V}}^{\mathfrak{U}}$ from $C^p(\mathfrak{V}, G)$ into $C^p(\mathfrak{U}, G)$:

$$\left(\tau_{\mathfrak{V}}^{\mathfrak{U}} c \right)_{i_0, i_1, \ldots, i_p} = c_{\tau(i_0), \tau(i_1), \ldots, \tau(i_p)}.$$

Thus we have

$$\tau_{\mathfrak{V}}^{\mathfrak{U}} \circ \delta = \delta \circ \tau_{\mathfrak{V}}^{\mathfrak{U}},$$

consequently, $\tau_{\mathfrak{V}}^{\mathfrak{U}}$ induces a mapping

$$\tau_{\mathfrak{V}}^{\mathfrak{U}} : H^p(\mathfrak{V}, G) \longrightarrow H^p(\mathfrak{U}, G).$$

The mapping $\tau_{\mathfrak{W}}^{\mathfrak{U}}$ is independent of the choice of τ (cf [36] for the proof of this assertion).

Now let $\mathfrak{U}, \mathfrak{V}, \mathfrak{W}$ be open coverings of M such that $\mathfrak{V} \prec \mathfrak{W} \prec \mathfrak{U}$. We have

$$\tau_{\mathfrak{V}}^{\mathfrak{W}} \tau_{\mathfrak{W}}^{\mathfrak{U}} = \tau_{\mathfrak{V}}^{\mathfrak{U}}.$$

We define $H^p(M, G)$ as follows. Consider the disjoint union of the groups $H^p(\mathfrak{U}, G)$. Let $u \in H^p(\mathfrak{U}, G)$ and $v \in H^p(\mathfrak{V}, G)$, then we say that

$$u \sim v$$

if there is an open covering $\mathfrak{W}$ refining $\mathfrak{U}$ and $\mathfrak{V}$ ($\mathfrak{U} \prec \mathfrak{W}$ and $\mathfrak{V} \prec \mathfrak{W}$) such that

$$\tau_{\mathfrak{V}}^{\mathfrak{W}}(v) = \tau_{\mathfrak{U}}^{\mathfrak{W}}(u).$$

Then, $u \sim v$ defines an equivalence relation. The set

$$H^p(M, G) = \varinjlim H^p(\mathfrak{U}, G)$$

of all equivalence classes is an abelian group with respect to the obvious operations. This group is called the *p-th Čech cohomology group of M* with coefficients in G.

Definition 9.1 *An open covering $\mathfrak{U} = (U_i)_{i \in I}$ of M is said to be contractible if, whenever $i_1 \ldots, i_p \in I$ and $U_{i_1} \cap \ldots \cap U_{i_p} \neq \emptyset$, the intersection $U_{i_1} \cap \ldots \cap U_{i_p}$ is contractible*

Recall that an open subset U of M is said to be contractible if the identity mapping $\mathrm{id}_U : U \longrightarrow U$ is homotopic to a constant mapping (that is, there exists a smooth mapping $\varphi : [0, 1] \times U \longrightarrow U$ such that $\varphi(0, x) = x$ for all $x \in U$ and $\varphi(0, x) = x_0$ where x_0 is a fixed point x_0 of $\mathfrak{U}$.

Let g be a Riemannian metric on M and let $\mathfrak{U} = (U_i)_{i \in I}$ be an open covering of M. For all $i \in I$ and $p \in U_i$ we consider a convex neighborhood $U_{p,i}$ of p in U. Thus we obtain a contractible refinement

$$(U_{p,i})_{i \in I, \ p \in U_i}$$

of $\mathfrak{U} = (U_i)_{i \in I}$ (for the definition and existence of convex neighborhoods, we can see [34]).

Theorem 9.1 *(Poincaré's lemma). Let U be a contractible open of M. Every closed p-form on U is exact.*

Theorem 9.2 *We have*

1. *The group $H^p(\mathfrak{U}, G)$ is isomorphic to $H^p(M, G)$ for every contractible covering $\mathfrak{U} = (U_i)_{i \in I}$ of M.*

2. *(de Rham's theorem) $H^p(M, \mathbb{K})$ is canonically isomorphic to $H_d^p(M, \mathbb{K})$ here $G = \mathbb{K}$ is the additive group $\mathbb{R}$ or $\mathbb{C}$.*

We have need of an explicit form of de Rham's theorem, hence we review in details the isomorphism for $p = 0, 1, 2$. For a complete proof (every p) we can refer to [58].

Proof. We give here the proof for $p = 0, 1, 2$. Let $\mathfrak{U} = (U_i)_{i \in I}$ be a contractible covering of M.

Let c be a 0-cochain $\big(c \in C^0(\mathfrak{U}, \mathbb{K})\big)$ such that $\delta c = 0$. We have:

$$(\delta c)_{i_0, i_1} = c_{i_0} - c_{i_1} = 0.$$

Thus if $U_{i_0} \cap U_{i_1} \neq \emptyset$ then $c_{i_0} = c_{i_1}$; consequently, the 0–cochain c defines a constant on M. Therefore $H^0(\mathfrak{U}, \mathbb{K}) \simeq \mathbb{K}$. Now, if f is a smooth function on M (that is, 0-form on M) such that $df = 0$, then f is constant on M (M is connected). Hence $H_d^0(M, \mathbb{K}) \simeq \mathbb{K}$. This proves the theorem for $p = 0$.

Now consider $p = 1$. Let ω be a closed $\mathbb{K}$-valued Pfaffian form on M. Since U_i is contractible for every $i \in I$, then by Poincaré's lemma there exists a smooth function $f_i : U_i \longrightarrow \mathbb{K}$ such that for each $i \in I$ we have

$$\omega_{|U_i} = df_i.$$

Thus if $U_i \cap U_j \neq \emptyset$ then we have

$$d\left(f_i - f_j\right) = 0$$

on $U_i \cap U_j$. Hence

$$(f_i - f_j)_{|U_i \cap U_j} = c_{i,j} \in \mathbb{K}.$$

We have associated with ω a 1-cochain $c \in C^1(\mathfrak{U}, \mathbb{K})$ such that $\delta c = 0$. Considering now an other family $g_i : U_i \longrightarrow \mathbb{K}$ such that for each $i \in I$ we have $\omega_{|U_i} = dg_i$, thus

$$(g_i - f_i)_{|U_i} = \lambda_i \in \mathbb{K}.$$

Set

$$b_{i,j} = (g_i - g_j)_{|U_i \cap U_j},$$

then we have

$$b_{i,j} = c_{i,j} + \lambda_i - \lambda_j = (c + \delta\lambda)_{i,j}.$$

Hence the equivalence class $[c]$ of c in $H^1(\mathfrak{U}, \mathbb{K})$ depends only on the closed differential form ω. It is clear that $[c] = 0$ if $\omega = df$. Therefore, we have a mapping

$$d.R : [\omega] \longmapsto [c]$$

from $H^1_d(M, \mathbb{K})$ into $H^1(\mathfrak{U}, \mathbb{K})$.

We now prove that this mapping is injective. Let ω be a closed Pfaffian form on M and let c be the corresponding 1−cochain. If $[c] = 0$ then there is $b \in C^0(\mathfrak{U}, \mathbb{K})$ such that

$$c_{i,j} = \delta b_{i,j} = b_i - b_j.$$

Let $h_i = f_i - b_i$. The restriction $(h_i - h_j)_{|U_i \cap U_j}$ of $h_i - h_j$ to $U_i \cap U_j$ is zero. Let h be the smooth function on M defined by

$$h(x) = h_i(x)$$

for every $x \in U_i$. We have $dh = \omega$, hence $[\omega] = 0$. This proves that the mapping $d.R$ is injective.

We must show that the mapping $d.R$ is surjective. Let $c \in C^1(\mathfrak{U}, \mathbb{K})$ with $\delta c = 0$. For a partition of unity $(\varphi_i)_{i \in I}$ subordinate to the covering $\mathfrak{U} = (U_i)_{i \in I}$ we consider

$$f_i = \sum_{j \in I(i)} \varphi_i c_{i,j},$$

where $I(i) = \{j \in I \mid U_i \cap U_j \neq \emptyset\}$.

From $\delta c = 0$ we have

$$(f_i - f_j)_{|U_i \cap U_j} = c_{i,j}.$$

For every $i \in I$ we take $\omega_i = df_i$. Then we have ω_i is a $\mathbb{K}$−valued 1-form on U_i and

$$(\omega_i - \omega_j)_{|U_i \cap U_j} = df_i - df_j = d(f_i - f_j) = 0.$$

Thus there exists a $\mathbb{K}$−valued 1-form ω on M, therefore

$$\omega_{|U_i} = \omega_i.$$

Now, following the above argument, we see that $[c]$ is the image under this mapping of $[\omega]$. This completes the proof in the case $p = 1$.

Suppose $p = 2$. Let ω be a $\mathbb{K}$−valued differential form of degree 2 such that $d\omega = 0$. By Poincaré's lemma there exists for every $i \in I$ a Pfaffian form α_i such that

$$\omega_{|U_i} = d\alpha_i.$$

For $U_i \cap U_j \neq \emptyset$ we have

$$d\left(\alpha_i - \alpha_j\right)_{|U_i \cap U_j} = \omega_{|U_i \cap U_j} - \omega_{|U_i \cap U_j} = 0,$$

hence there exists a smooth function $f_{ij} : U_i \cap U_j \longrightarrow \mathbb{K}$ such that

$$\alpha_i - \alpha_j = df_{ij}.$$

Now suppose $U_i \cap U_j \cap U_k \neq \emptyset$. Then we have

$$d\left(f_{ij} - f_{ik} + f_{jk}\right) = \alpha_i - \alpha_j - \alpha_i + \alpha_k + \alpha_j - \alpha_k = 0.$$

Let

$$c_{i,j,k} = \left(f_{ij} - f_{ik} + f_{jk}\right)_{|U_i \cap U_j \cap U_k}.$$

We have $c_{i,j,k} \in \mathbb{K}$. It is clear that $c \in C^2(\mathfrak{U}, \mathbb{K})$ and $\delta c = 0$.

Arguing as in the case $p = 1$, we see that the association $\omega \longmapsto [c]$ is well defined and induces a mapping from $H^2_d(M, \mathbb{K})$ into $H^2(\mathfrak{U}, \mathbb{K})$.

Let ω be a closed $\mathbb{K}$-valued 2-form on M such that the corresponding 2-cochain $c \in C^2(\mathfrak{U}, \mathbb{K})$ satisfies $c = db$; we must show that ω is an exact 2-form on M. In the previous notations let $\overline{f}_{ij} : U_i \cap U_j \longrightarrow \mathbb{K}$ be the mapping defined by

$$\overline{f}_{ij} = f_{ij} - b_{ij}.$$

We have

$$\overline{f}_{ij} - \overline{f}_{ik} + \overline{f}_{jk} = 0$$

on $U_i \cap U_j \cap U_k$. Let $(\varphi_i)_{i \in I}$ be a partition of unity subordinate to the covering $\mathfrak{U} = (U_i)_{i \in I}$, and let

$$f_i = \sum_{j \in I(i)} \varphi_j f_{ij},$$

where $I(i) = \{j \in I \mid U_i \cap U_j \neq \emptyset\}$. It is not difficult to see that we have

$$(f_i - f_j)_{|U_i \cap U_j} = \overline{f}_{ij}.$$

Let

$$\overline{\alpha}_i = \alpha_i - df_i.$$

On $U_i \cap U_j$, we have

$$\overline{\alpha}_i = \overline{\alpha}_j,$$

consequently there exists a $\mathbb{K}$-valued Pfaffian form $\overline{\alpha}$ defined on M such that the restriction of $\overline{\alpha}$ to U_i coincides with $\overline{\alpha}_i$ (for every $i \in I$):

$$\overline{\alpha}_{|U_i} = \overline{\alpha}_i.$$

It is clear that we have $\omega = d\bar{\alpha}$, thus the mapping $[\omega] \longmapsto [c]$ from $H_d^2(M, \mathbb{K})$ into $H^2(\mathfrak{U}, \mathbb{K})$ is injective. To complete the proof in the case $p = 2$ we must show that this mapping is surjective. Let $c \in C^1(\mathfrak{U}, \mathbb{K})$ with $\delta c = 0$. We show that there exists a closed $\mathbb{K}$-valued two form ω on M, thus:

1. $\omega_{|U_i} = d\alpha_i$;

2. $(\alpha_i - \alpha_j)_{|U_i \cap U_j} = df_{ij}$;

3. $f_{ij} - f_{ik} + f_{jk} = c_{i,j,k}$. on $U_i \cap U_j \cap U_k$.

Let $(\varphi_i)_{i \in I}$ be a partition of unity subordinate to the covering $\mathfrak{U} = (U_i)_{i \in I}$ and let $f_{ij} : U_i \cap U_j \longrightarrow \mathbb{K}$ be the mapping defined by

$$f_{ij} = \sum_{k \in I} (\varphi_k)_{|U_i \cap U_j \cap U_k} c_{i,j,k}.$$

Let

$$\alpha_i = \sum_{j \in I} \varphi_j \, (df_{ij})_{|U_i \cap U_j}$$

and

$$\omega_i = d\alpha_i$$

for every $i \in I$. Then we have

$$(\omega_i - \omega_j)_{|U_i \cap U_j} = 0.$$

Thus there exists a $\mathbb{K}$-valued differential form ω on M of degree 2 such that

$$\omega_{|U_i} = \omega_i$$

for all $i \in I$. It is clear that the corresponding element to ω under this mapping is $[c]$. This completes the proof in the case $p = 2$ and we have shown both parts of this theorem for $p = 0, 1, 2$. $\blacksquare$

9.2 Line bundle

Definition 9.2 *A line bundle is a triplet (L, p, M), where E and M are smooth manifolds and $p : E \longrightarrow M$ is a smooth mapping satisfying the following conditions:*

1. *For every $x \in M$, the fibre $L_x = p^{-1}(x)$ over x is a complex vector space of dimension 1;*

2. *There exists an open covering $(U_i)_{i \in I}$ of M and a family of diffeomorphisms*

$$\varphi_i : p^{-1}(U_i) \longrightarrow U_i \times \mathbb{C}$$

such that, for every $x \in U_i$ and $y \in L_x$ we have:

$$\varphi_i(y) = (x, \varphi_{i,x}(y))$$

and

$$\varphi_{i,x} : L_x \longrightarrow \mathbb{C}$$

is $\mathbb{C}-$linear.

Proposition 9.1 *Let (L, p, M) be a line bundle and let $\mathfrak{U} = (U_i)_{i \in I}$ be a contractible covering of M. Then there exists a family of diffeomorphisms $\varphi_i : p^{-1}(U_i) \longrightarrow U_i \times \mathbb{C}$ such that for every $x \in U_i$ and $y \in L_x$ we have*

$$\varphi_i(y) = (x, \varphi_{i,x}(y))$$

and

$$\varphi_{i,x} : L_x \longrightarrow \mathbb{C}$$

is $\mathbb{C}-$linear. Let $i, j \in I$ be such that $U_i \cap U_j \neq \emptyset$ then the mapping

$$\varphi_i \circ \varphi_j^{-1} : U_i \cap U_j \times \mathbb{C} \longrightarrow U_i \cap U_j \times \mathbb{C}$$

satisfies

$$\varphi_i \circ \varphi_j^{-1}(x, z) = (x, g_{ij}(x)(z))$$

for all $x \in U_i \cap U_j$ and $z \in \mathbb{C}$. Therefore

$$g_{ij} : U_i \cap U_j \longrightarrow \mathbb{C} - \{0\}$$

is C^∞.

The mappings g_{ij} satisfy the condition

$$g_{ij} g_{jk} = g_{ik} \tag{9.1}$$

on $U_i \cap U_j \cap U_k$.

Remark 14 *Observe that condition 9.1 is $\delta g = 0$ written multiplicatively.*

But $U_i \cap U_j$ is contractible, thus there is a smooth mapping

$$h_{ij} : U_i \cap U_j \longrightarrow \mathbb{C},$$

such that

$$g_{ij} = \exp(2i\pi h_{ij}).$$

The relation 9.1 implies that the mappings h_{ij} satisfy

$$h_{ij} + h_{jk} - h_{ik} \equiv c_{i,j\ ,k} \in \mathbb{Z}$$

for all $x \in U_i \cap U_j \cap U_k$ and $i, j, k \in I$.
It is clear that $c \in C^2(\mathfrak{U}, \mathbb{Z})$ and $\delta c = 0$.

The class of c in $H^2(\mathfrak{U}, \mathbb{Z})$ is independent of the choices made in its definition, the class of c is denoted by $K(L)$ and is called the characteristic (or Chern) class of L. The following properties are equivalent:

1. (L, p, M) is trivial (that is, isomorphic with $M \times \mathbb{C}$);

2. $K(L) = 0$.

Let $c \in C^2(\mathfrak{U}, \mathbb{Z})$ and $\delta c = 0$, then $c \in C^2(\mathfrak{U}, \mathbb{K})$ where $\mathbb{K} = \mathbb{R}$ or $\mathbb{C}$ with $\delta c = 0$. If $c = \delta b$ with $b \in C^{k-1}(\mathfrak{U}, \mathbb{Z})$ then $b \in C^{k-1}(\mathfrak{U}, \mathbb{Z})$. Thus we have a mapping

$$i : H^k(\mathfrak{U}, \mathbb{Z}) \longrightarrow H^k(\mathfrak{U}, \mathbb{K}),$$

where $\mathbb{K} = \mathbb{R}$ or $\mathbb{C}$.

Let $\varsigma \in i\left(H^2(M, \mathbb{K})\right)$ where $\mathbb{K} = \mathbb{R}$ or $\mathbb{C}$. We say that ς is integral if $\varsigma \in i\left(H^2(M, \mathbb{Z})\right)$.

Let $\theta \in \Lambda^2(M, \mathbb{K})$ be a $\mathbb{K}$-valued differential form of degree 2 on M. We say that θ is integral if, the image of $[\theta]$ by the de Rham isomorphism, is integral.

9.3 Connections

A cross-section of a line bundle (L, p, M) over M is a smooth mapping $\sigma : M \longrightarrow L$ such that $p \circ \sigma = id_M$.

We denote by $\Gamma(L)$ the space of all cross-sections of the line bundle (L, p, M).

A connection on L is a $\mathbb{C}$-bilinear pairing for each $x \in M$,

$$(X_x, \sigma) \longmapsto \nabla_{X_x} \sigma$$

of $T_x M$ with $\Gamma(L)$ into the fibre L_x such that

1. $\nabla_{X_x} f\sigma = X_x(f)\sigma + f(x)\,\nabla_{X_x}\,\sigma$, for all $X_x \in T_x M$, $\sigma \in \Gamma(L)$ and $f \in C^\infty(M, \mathbb{C})$;

2. For all $X \in \mathfrak{X}(M)$, $f \in C^\infty(M, \mathbb{C})$ and $\sigma \in \Gamma(L)$, the association

$$\nabla_X \sigma : x \longmapsto \nabla_{X_x} \sigma$$

   defines a smooth mapping from M into $\Gamma(L)$, here $C^\infty(M, \mathbb{C})$ denotes the set of all smooth mappings from M into $\mathbb{C}$.

Proposition 9.2 *Each line bundle admits a connection.*

Proof. Let (L, p, M) be a line bundle over M and let $\mathfrak{U} = (U_i)_{i \in I}$ be an open covering on M and $(\varphi_i)_{i \in I}$ be the family of diffeomorphisms

$$\varphi_i : p^{-1}(U_i) \longrightarrow U_i \times \mathbb{C}$$

such that for every $x \in U_i$ and $y \in L_x$ we have

$$\varphi_i(y) = (x, \varphi_{i,x}(y))$$

and

$$\varphi_{i,x} : L_x \longrightarrow \mathbb{C}$$

is $\mathbb{C}$-linear. For each $i \in I$ we take

$$\sigma_i(x) = \varphi_i^{-1}(x, 1)$$

for every $x \in U_i$. Let $\sigma \in \Gamma(L)$ we have

$$\sigma(x) = h_i(x)\sigma_i(x)$$

for every $x \in U_i$, where $h_i : U_i \longrightarrow \mathbb{C}$ is a smooth mapping.
    For each $i \in I$ we take

$$\nabla^i_{X_x} \sigma = X_x(h_i)\sigma_i(x)$$

for all $x \in U_i$ and $X_x \in T_x M$. Let $(\theta_i)_{i \in I}$ be a partition of unity subordinate to the covering $\mathfrak{U}$. Define

$$\nabla_{X_x} \sigma = \sum_{i \in I} \theta_i(x)\,\nabla^i_{X_x}\,\sigma$$

for all $X_x \in TM$ and $\sigma \in \Gamma(L)$.
It is not difficult to see that ∇ defines a connection on L. ∎

Let (L, p, M) be a line bundle over M equipped with a connection ∇, let $\mathfrak{U} = (U_i)_{i \in I}$ be an open covering on M and let $(\varphi_i)_{i \in I}$ be the family of diffeomorphisms

$$\varphi_i : p^{-1}(U_i) \longrightarrow U_i \times \mathbb{C}$$

such that, for every $x \in U_i$ and $y \in L_x$, we have:

$$\varphi_i(y) = (x, \varphi_{i,x}(y))$$

and

$$\varphi_{i,x} : L_x \longrightarrow \mathbb{C}$$

is $\mathbb{C}$-linear. For each $i \in I$ we define

$$\sigma_i(x) = \varphi_i(x, 1)$$

for every $x \in U_i$. For each $\sigma \in \Gamma(E)$ we have

$$\sigma_{|U_i} = f_i \sigma_i,$$

where $f_i : U_i \longrightarrow \mathbb{C}$ is a smooth function on U_i. Thus

$$\nabla_{X_x} \sigma = X_x(f_i)\sigma_i(x) + f_i(x) \nabla_{X_x} \sigma_i$$

for each $x \in U_i$. The connection is specified by $\nabla_{X_x}\sigma_i$. But $\sigma_i(x) \neq 0$ for every $x \in U_i$, hence

$$\nabla_{X_x}\sigma_i = \theta_i(X_x)\sigma_i$$

for $x \in U_i$ and $X_x \in T_x M$.

It is clear that θ_i is a $\mathbb{C}$-valued differential form of degree 1 on U_i. For each $x \in U_i \cap U_j$ we have

$$\sigma_i(x) = g_{ij}(x)\sigma_j(x).$$

Hence

$$\nabla_{X_x}\sigma_i = X_x(g_{ij})\sigma_j + g_{ij} \nabla_{X_x} \sigma_j$$

for $x \in U_i \cap U_j$ and $X_x \in T_x M$. Therefore we have

$$\theta_i(X_x) = X_x(g_{ij})g_{ij}^{-1} + \theta_j(X_x)$$

for $x \in U_i \cap U_j$ and $X_x \in T_x M$; that is,

$$\theta_i = g_{ij}^{-1} dg_{ij} + \theta_j; \tag{9.2}$$

But, $d\left(g_{ij}^{-1}dg_{ij}\right) = 0$ on $U_i \cap U_j$, thus

$$d\theta_i = d\theta_j$$

on $U_i \cap U_j$. This proves that there exists a $\mathbb{C}$−valued differential form ω of degree 2 on M whose the restriction to U_i coincides with $d\theta_i$

$$\omega_{|U_i} = d\theta_i$$

for every $i \in I$.

Proposition 9.3 *The 2-form ω is independent of all choices used to define it and we have:*

$$\omega(X,Y)\sigma = \nabla_X \nabla_Y \sigma - \nabla_Y \nabla_X \sigma - \nabla_{[X,Y]}\sigma$$

for all $\sigma \in \Gamma(L)$ and $X,Y \in \mathfrak{X}(M)$.

Proof. It is clearly enough to prove the formula in the statement of the proposition. First we have:

$$\left(\nabla_X \nabla_Y - \nabla_Y \nabla_X - \nabla_{[X,Y]}\right)(f\sigma) = f\left(\nabla_X \nabla_Y \sigma - \nabla_Y \nabla_X \sigma - \nabla_{[X,Y]}\sigma\right) \tag{9.3}$$

for every smooth mapping $f : M \longrightarrow \mathbb{C}$. Hence it is enough to prove the formula for $\sigma = \sigma_i$ on U_i.

$$\nabla_X \nabla_Y \sigma_i \quad - \nabla_Y \nabla_X \sigma_i - \nabla_{[X,Y]}\sigma_i$$
$$= \nabla_X \left(\theta_i(Y)\sigma_i\right) - \nabla_Y \left(\theta_i(X)\sigma_i\right) - \theta_i \left([X,Y]\right)\sigma_i$$
$$= X\left(\theta_i(Y)\right)\sigma_i + \theta_i(Y)\theta_i(X)\sigma_i - Y\left(\theta_i(X)\right)\sigma_i - \theta_i(X)\theta_i(Y)\sigma_i$$
$$\quad -\theta_i\left([X,Y]\right)\sigma_i$$
$$= \left(X\left(\theta_i(Y)\right) - Y\left(\theta_i(X)\right) - \theta_i\left([X,Y]\right)\right)\sigma_i.$$

The proposition results from the property

$$d\theta(X,Y) = X\left(\theta(Y)\right) - Y\left(\theta(X)\right) - \theta\left([X,Y]\right)$$

for any $\mathbb{C}$−valued differential one form θ and $X,Y \in \mathfrak{X}(M)$. ∎

Definition 9.3 *The two form ω is called the curvature of L associated with ∇ and denoted by $\mathrm{curv}(L,\nabla)$.*

It is clear that we have

$$d\left(\mathrm{curv}(L,\nabla)\right) = 0.$$

Proposition 9.4 *We have:*

$$\frac{1}{2i\pi}d.R(curv(L,\nabla)) = i(K(L)),$$

where d.R is the de Rham isomorphism and $K(L)$ is the Chern class of L. In particular

$$\frac{1}{2i\pi}d.R(\mathrm{curv}(L,\nabla))$$

is an integral form.

Proof. Let $\mathfrak{U} = (U_i)_{i \in I}$ be a contractible covering of M. From the discussion above we have

$$\omega_{|U_i} = d\theta_i$$

and

$$(\theta_i - \theta_j) = g_{ij}^{-1}dg_{ij} = d\left(\log g_{ij}\right),$$

thus

$$\frac{1}{2i\pi}\left(\log g_{ij} - \log g_{ik} + \log g_{jk}\right) = c_{i,j,k}$$

on $U_i \cap U_j \cap U_k$. It is clear that we have $[c] \in H^2(M,\mathbb{C})$ and $[c] = i(K(L))$. Which proves the proposition. ∎

Definition 9.4 *Let L be a line bundle over M. A Hermitian structure on L is a correspondence $x \longmapsto \langle,\rangle_x$, where $\langle,\rangle_x$ is a positive-definite inner product on L_x such that for any $\sigma \in L$ the mapping*

$$x \longmapsto \langle \sigma(x), \sigma(x)\rangle_x$$

is smooth on M.

A Hermitian line bundle is a pair $(L, \langle,\rangle)$ consisting of a line bundle and a Hermitian structure on L.

Definition 9.5 *Let L be a line bundle over M endowed with a connection ∇ and a Hermitian structure $\langle,\rangle$. We say that the Hermitian structure $\langle,\rangle$ is ∇-invariant if*

$$X\left(\langle\sigma_1,\sigma_2\rangle\right) = \langle\nabla_X\sigma_1,\sigma_2\rangle + \langle\sigma_1,\nabla_X\sigma_2\rangle$$

for all $X \in \mathfrak{X}(M)$ and $\sigma_1,\sigma_2 \in \Gamma(L)$.

Proposition 9.5 *Let L be a line bundle equipped with a connection ∇.*

1. *There is at most one (up to scalar multiple) ∇-invariant Hermitian structure on L.*

2. *If there exists a ∇-invariant Hermitian structure on L then*

$$\frac{1}{2i\pi}\operatorname{curv}(L,\nabla)$$

   is an $\mathbb{R}$-valued differential 2-form.

3. *If $H^1(M,\mathbb{R})$ is trivial then there is a ∇-invariant Hermitian structure on L if and only if*

$$\frac{1}{2i\pi}\operatorname{curv}(L,\nabla)$$

   is an $\mathbb{R}$-valued differential 2-form.

Proposition 9.6 *Let η be an $\mathbb{R}$-valued differential form of degree 2 on M such that $d.R([\eta]) \in i\left(H^2(M,\mathbb{Z})\right)$. Then there exists a line bundle $(L(\eta),p,M)$ over M equipped with a connection $\nabla(\eta)$ such that*

$$\frac{1}{2i\pi}\operatorname{curv}(L(\eta),\nabla(\eta)) = \eta.$$

Furthermore, if $H^1(M,\mathbb{Z})$ is trivial then M admits a ∇-invariant Hermitian structure.

   Let $L^* = \{l \in L \mid l \neq 0\}$.

Proposition 9.7 *Let (L,p,M) be a line bundle equipped with a connection ∇. Then there exists a unique $\mathbb{C}$-valued differential form θ of degree 1 on L^* $\left(\theta \in \Lambda^1(L^*,\mathbb{C})\right)$ such that if $U \subset M$ is an open subset and a non-vanishing cross section σ over U $\left(\sigma \in \Gamma(L_{|U})\right)$ then*

$$\nabla_{X_x}\sigma = (\sigma^*\theta)(X_x)\sigma(x)$$

for $x \in M$ and $X_x \in T_x M$. θ is called the connection form of ∇.

9.4 k-symplectic pre-quantization

Let M be an $n(k+1)$-dimensional manifold equipped with a k-symplectic structure

$$(\theta^1, \ldots, \theta^k; E).$$

Definition 9.6 *([49]) We say that M is a quantizable manifold (with respect to the k−symplectic structure) if*

$$\frac{[\theta^1]}{2i\pi}, \ldots, \frac{[\theta^k]}{2i\pi}$$

represent integral cohomology classes of M.

Example 26 *Let M be an $n(k+1)$-dimensional manifold equipped with a k-symplectic structure $(\theta^1, \ldots, \theta^k; E)$ such that the two forms $\theta^1, \ldots, \theta^k$ are exact, then M is a quantizable manifold.*

Definition 9.7 *Let M be a quantizable manifold. Then there exist line bundles*

$$\left(L^1, \nabla^1\right), \ldots, \left(L^k, \nabla^k\right)$$

such that

$$\operatorname{curv}\left(L^p, \nabla^p\right) = \theta^p$$

for each $p = 1, \ldots, k$. The bundle $L = L^1 \oplus \ldots \oplus L^k$ with the connection ∇ given by

$$\nabla_X \left(\sigma^1 \oplus \ldots \oplus \sigma^k\right) = \nabla_X^1 \sigma^1 \oplus \ldots \oplus \nabla_X^k \sigma^k$$

is called the pre-quantum bundle of M. The Hermitian structure on L is given in a natural way via the Hermitian structure on each (L^p, ∇^p), $p = 1, \ldots, k$.

Example 27 *Let M be an $n(k+1)$-dimensional manifold equipped with a k-symplectic structure $(\theta^1, \ldots, \theta^k; E)$ such that $\theta^p = d\omega^p$, for each $p = 1, \ldots, k$. Then M is a quantizable manifold and for each $p = 1, \ldots, k$ we have*

1. $L^p = M \times \mathbb{C}$;

2. $\nabla_X^p \sigma = X(\sigma) - i\left(i(X)\omega^p\right)$;

3. $\langle (x, z_1), (x, z_1) \rangle_{L^p} = z_1 \overline{z_2}$.

Let M be a quantizable manifold with respect to the k-symplectic structure $(\theta^1, \ldots, \theta^k; E)$, let (L, ∇) be the corresponding pre-quantum bundle and $\Gamma_c(L^p)$ be the space of smooth sections of L^p with compact support. We denote by $\mathcal{H}^p$ the completion of $\Gamma_c(L^p)$ with respect to the inner product

$$\langle \sigma_1, \sigma_2 \rangle_{L^p} = \int (\sigma_1, \sigma_2)$$

where $(\sigma_1, \sigma_2) : x \in M \longmapsto (\sigma_1(x), \sigma_2(x)) \in \mathbb{C}$. Then the pre-quatum Hilbert representation space is

$$\mathcal{H} = \mathcal{H}^1 + \ldots + \mathcal{H}^k.$$

If we choose $H \in \mathfrak{H}(M)$, $H = (H^1, \ldots, H^k)$ then its representation as an operator δ_H on $\mathcal{H}$ can be defined by:

$$\delta_H : \sigma = (\sigma_1, \ldots, \sigma_k) \in \mathcal{H} \longmapsto \delta_H(\sigma) \in \mathcal{H}$$

with

$$\delta_H(\sigma) = (\delta^1_{H^1}(\sigma_1), \ldots, \delta^k_{H^k}(\sigma_k)),$$

where for each $p = 1, \ldots, k$

$$\delta^p_{H^p}(\sigma_p) = -i\hbar \nabla^p_{X_H} \sigma_p + H^p \sigma_p.$$

Proposition 9.8 *For all* $H, K \in \mathfrak{H}(M)$ *we have:*

$$[\delta_H, \delta_K] = i\hbar \delta_{\{H, K\}}.$$

As consequence we have:

Proposition 9.9 *The correspondence*

$$H \in \mathfrak{H}(M) \longmapsto -\frac{1}{i\hbar} \delta_H : \mathcal{H} \longrightarrow \mathcal{H}$$

is a representation of the Lie algebra $\mathfrak{H}(M)$ *by skew-Hermitian first order differential operators.*

Bibliography

[1] E. ARTIN *Algèbre géométrique* Gauthiers - Villars (1972).

[2] L. AUSLANDER Examples of locally affine spaces *Ann. of Maths.* **64** (1964) 255 - 259.

[3] L. AUSLANDER The structure of complete locally affine manifolds. *Topology* **3** (Suppl.1) (1964) 131 - 139.

[4] L. AUSLANDER and L. MARKUS Holonomy of flat affinely connected manifolds. *Ann. of Math.* Princeton **62** (1955) 139 - 151.

[5] A. AWANE Sur une généralisation des structures symplectiques. Thesis, Strasbourg (1984).

[6] A. AWANE *k*-symplectic structures . *Journal of Mathematical Physics* **33** (1992) 4046 - 4052. U.S.A.

[7] A. AWANE G-espaces *k*-symplectiques homogènes. *Journal of Geometry and Physics* **13** (1994) 139 - 157. North-Holland.

[8] A. AWANE Structures *k*-symplectiques. Thesis, Mulhouse(1992).

[9] A. AWANE Some affine properties of the *k*-symplectic manifolds. *Beiträge zur Algebra und Geometrie* , *Contribution to Algebra and Geometry* **39** (1998) , 75 - 83. Germany

[10] A. AWANE Systèmes extérieures *k*-symplectiques. To appear in Rend. Sem. Mat. Univers. Politecn. Torino.

[11] ROBERT A. BLUMENTHAL Foliated manifolds with flat basic connection. *J. Differential Geometry* , **16** (1981) 401 - 406.

[12] R. BOTT *Lectures on characteristic classes and foliations.* Lecture note in Mathematics, 279 (Springer-Verlag, New-York . 1972) 1 - 80.

[13] R. BRYANT, SS. CHERN, R. GARDNER, H. GOLDSCHMIDT, P. GRIFFITH *Exterior Differential Systems.* MSRI Publications 18. Springer Verlag (1991).

[14] E. CARTAN Les systèmes de Pfaff à 5 variables. *Ann. Sci. Ec. Nor. Sup.* **27**(1910) 109-192.

[15] L. CONLON Transversally parallelizable foliations of codimension 2. *Trans. Amer. Math. soc.* **194** (1974) 79 - 102.

[16] P. DAZORD Sur la géométrie des sous-fibrés et des feuilletages lagrangiens. *Ann. Ecole Normale Sup.* **14** (1981) 465 - 480.

[17] J. DIEUDONNE *La géométrie des groupes classiques.* Springer-Verlag (1971).

[18] J. DIEUDONNE *Eléments d'Analyse.* Gauthiers-Villars (1974).

[19] J. DIEUDONNE *Linear Algebra and Geometry.* Hermann, Paris (1969)

[20] R. GARDNER Invariants of Pfaffian systems. *T.R.A.M.S.* **126** (1967) 514 - 533.

[21] M. GEOFFREY Dynamical structures for k-vectors fields. *International Journal of Theorical Physics.* **27**, 5, (1988) 571 - 585.

[22] M. GEOFFREY A Darboux theorem for multi-symplectic manifolds. *Lectures in Mathematical Physics.* **16**, (1988) 133 - 138.

[23] C. GODBILLON *Géométrie différentielle et Mécanique Analytique.* Hermann. Paris (1969).

[24] C. GODBILLON *FEUILETAGES. Etude géométrique.* Birkhüser (1991).

[25] C. GODBILLON *Eléments de topologie algébrique.* Hermann. Paris (1971).

[26] H. GOLDSCMIDT, S. STERNBERG. The Hamiltonian Cartan formalism in the calcul of variations. *Ann. Inst. Fourier.* **23**, (1973), 203 - 267.

[27] E. GOURSAT *Leçons sur les systèmes de Pfaff.* Paris. 1922.

[28] M. GOZE Systèmes de Pfaff. *Rendiconti Seminario. Facultà di Scienze. Università di Cagliari,* **60** Fasc.2 (1990) 167 - 187.

[29] M. GOZE Systèmes de Pfaff associés aux algèbres de type H. *Rend. Sem. Mat. Univers. Politecn. Torino* , **46**, 1 (1988) 91 - 110.

[30] M. GOZE Sur la classe des formes invariantes à gauche sur un groupe de Lie *CRAS, Paris*, **283**, (1976), 499-502.

[31] M. GOZE - A. BOUYAKOUB Sur les algèbres de Lie munie d'une forme symplectique. *Rendiconti Seminario Facoltà Scienze. Università Calgiari* , **37** ,1(1987) 86 - 97.

[32] M. GOZE - Y. HARAGUCHI Sur les r-systèmes de contact. *CRAS, Paris*, **294**, (1982), 95 - 97.

[33] M. GOZE - Y. KHAKIMDJANOV *Nilpotent Lie algebras*. Kluwer Academic Publishers. Dordreicht / Boston / London (1996).

[34] S. HELGASON *Differential Geometry and Symmetric spaces* . Academic Press. New-York (1978).

[35] R. HERMANN *Lie groups for Physicists*. W.A. Benjamin. New York (1966)

[36] F. HIRZBRUCH *Topological methods in Algebraic Geometry*. Springer Verlag, New York (1968).

[37] KIJOWSKI JERRAY WLODZIMIERS TULEZYJEW *A symplectic framework for field theories*. Lectures Notes in Physics, Vol 107, Springer Verlag, Berlin, (1979).

[38] S. KIJOWSKI, W. SZCZYRBA Multisymplectic manifolds and Lie geometric construction on the Poisson bracket in field theory. *Géométrie symplectique et Physique mathématique. CNRS Paris*, (1975), 347 - 378.

[39] S. KOBAYASHI and K. NOMIZU *Foundations of differential Geometry*. Volume 1. Interscience Publishers New-York (1963).

[40] B. KOSTANT *Quantization and representation theory*. Lecture notes in Math. Vol 170, Springer Verlag Berlin (1970).

[41] M. de LEON-MENDEZ-SALGADO Regular p-almost cotangent structures. *J. Corean Math. Soc.* **25,** 2, (1988) 273 - 287.

[42] P. LIBERMANN et C.M. MARLE *Géométrie symplectique Bases théorique de la Mécanique classique*. Tomes 1, 2, 3, 4 U.E.R. de Mathématiques, L.A. 212 et E.R.A. 944, 1020, 1021 du C.N.R.S.

[43] A. MEDINA *Structures de Poisson affines.* Colloque international Aix-en-Provence (1990).

[44] P. MOLINO *Géométrie de Polarisation.* Travaux en cours Hermann (1984) 37-53.

[45] P. MOLINO Géométrie globale des feuilletages riemanniens. *Proc. Kon. Nederl. Akad.* Ser.A, 1, **85** (1982) 45 - 76.

[46] K. NOMIZU *Lie Groups and differential Geometry.* Math. Soc. Japan (1956).

[47] Y. NAMBUGeneralized Hamiltonian Dynamics. *Physical Review* D , **7**, 8 (15 April 1973).

[48] M. POSTNIKOV *Groupes et Algèbres de Lie.* Mir. Moscou.(1985).

[49] M. PUTA Some Remarks on the k-symplectic manifolds. *Tensors.*109 - 115.

[50] T. SARI Sur les variétés de contact localement affines. *CRAS, Paris,* **292**, (1981), 809 - 812.

[51] T. SARI Sur les variétés de contact affines plates. *Séminaire Gaston-Darboux de Géométrie et Topologie,* (1991-1992).

[52] S. STERNBERG *Lectures on differential Geometry.* Prentice Hall (19964).

[53] S. SZAPIRO Geodesic fields in the calcul of variations of multiple integrals depending on derivatives of higher order. *Lecture notes in math.* **836**, (1980), 501-511.

[54] N. WALLACH *Symplectic Geometry and Fourier Analysis.* MATH. SCI. PRESS, 53, Jordan Road, Brookline, Massachusetts. 02146 (1977).

[55] G. WARNER *Foundations of differential manifolds and Lie groups.* Scott, Foresman and co. Glenview (1972).

[56] A. WEINSTEIN *Lectures on symplectic manifolds.* Conference board of mathematical science, (Regional Conference Series in Mathematics n 29 , A.M.S.) (1977).

[57] A. WEINSTEIN Symplectic manifolds and their Lagrangian submanifolds. *Advances in Maths,* **6**, (1971), 329 - 346.

[58] A. WIEL Sur le théoràme de de Rham. *Comm. math. Helv.* **26,** (1952), 119 - 145.

[59] J.A. WOLF *Spaces of constant curvature* (University of California, 1972).

Index